## *Praise for*
## Presentations in Action

"Jerry Weissman is the Jedi Master of presentations and effective communication. *Presentations in Action* is a wonderful compilation of 80 interesting examples and stories that will make you think and help you improve your presentations and public speaking. I've added this important and to-the-point book to my Jerry Weissman collection. Another fabulous read from the Silicon Valley legend."

—**Garr Reynolds**, author of *Presentation Zen*
and *The Naked Presenter*

"Across all fields there is one common trait of leaders: the ability to persuade groups to follow. This is the field guide to persuasion, thus the field guide to successful leadership."

—**Scott Cook**, Co-Founder and Chairman
of the Executive Committee, Intuit, Inc.

"The best way to learn how to become great is to study the greatest. In *Presentations in Action*, the world's #1 presentations consultant presents 80 succinct lessons explaining what made the masters effective. These simple lessons make it easier to be much better. Jerry taught me how to capture my audience's attention in the first minute of my talks with just six words, 'Tell a story, not a joke.' What a difference."

—**Bill Davidow**, Venture Capitalist, author of *Marketing High Technology* and *Overconnected*

"Jerry gives you 80 secrets from the world's best persuaders, compacted into bite-sized chapters that make them easy to read and easy to apply. Taken together they define the dynamics of communication that can and have changed the world."

—**Peter Guber**, Chairman of Mandalay Entertainment

"There's nothing I love more than case studies and great presentations. Jerry's book provides case studies so you can make great presentations. It doesn't get more enchanting than this."

—**Guy Kawasaki**, author of *Enchantment:
The Art of Changing Hearts, Minds, and Actions*

"Loved the book; it is the key to 'message received' because what is said is less important than what is heard!"

—**Vinod Khosla**, Partner, Khosla Ventures

*"Eighty Presentations Ideas from the Masters* is like listening to football advice from Vince Lombardi—elegant, purposeful, and direct. In this compact but complete book, Jerry Weissman gives you all the right slants on public speaking and presentations. My favorite is Chapter 46, advice from Frank Sinatra—who puts lyrics ahead of melody. Congratulations, Jerry, on an insightful masterwork."

—**C. Richard Kramlich**, Chairman, New Enterprise Associates

"Jerry's coaching has been invaluable for many of our c-suite clients as they prepare for their debut or return to the public markets. His high-impact presentation approach has been tremendously successful. In addition, Jerry's book series that detail his differentiating concepts have been incredibly instructive."

—**Michael Millman**, Managing Director, J.P. Morgan–Equity Capital Markets

# Presentations in Action

# Presentations in Action:

*80 Memorable Presentation Lessons from the Masters*

Jerry Weissman

Vice President, Publisher: Tim Moore
Associate Publisher and Director of Marketing: Amy Neidlinger
Executive Editor: Jeanne Glasser
Editorial Assistant: Pamela Boland
Art Consultant: Nichole Nears
Development Editor: Russ Hall
Operations Manager: Gina Kanouse
Senior Marketing Manager: Julie Phifer
Publicity Manager: Laura Czaja
Assistant Marketing Manager: Megan Colvin
Cover Designer: Alan Clements
Managing Editor: Kristy Hart
Senior Project Editor: Lori Lyons
Copy Editor: Krista Hansing Editorial Services, Inc.
Proofreader: Chrissy White, Language Logistics, LLC
Senior Indexer: Cheryl Lenser
Senior Compositor: Gloria Schurick
Manufacturing Buyer: Dan Uhrig

Pearson Education, Inc.
Publishing as FT Press
Upper Saddle River, New Jersey 07458

FT Press offers excellent discounts on this book when ordered in quantity for bulk purchases or special sales. For more information, please contact U.S. Corporate and Government Sales, 1-800-382-3419, corpsales@pearsontechgroup.com. For sales outside the U.S., please contact International Sales at international@pearson.com.

Printed in the United States of America
Second Printing June 2011

Pearson Education LTD.
Pearson Education Australia PTY, Limited.
Pearson Education Singapore, Pte. Ltd.
Pearson Education Asia, Ltd.
Pearson Education Canada, Ltd.
Pearson Educación de Mexico, S.A. de C.V.
Pearson Education—Japan
Pearson Education Malaysia, Pte. Ltd.

*Library of Congress Cataloging-in-Publication Data*

Weissman, Jerry.
  Presentations in action : 80 memorable presentation lessons from the masters / Jerry Weissman.
    p. cm.
  Includes bibliographical references.
  ISBN 978-0-13-248962-1 (hardback : alk. paper) 1. Business presentations. I. Title.
  HF5718.22.W4495 2011
  658.4'52—dc22
                                                            2010050905

ISBN-10: 0-13-248962-7
ISBN-13: 978-0-13-248962-1

For my Lovely Lady Lucie
again . . . and again

# Contents

Introduction . . . . . . . . . . . . . . . . . . . . . . . .1

**SECTION I**   **Content: The Art of Telling Your Story**

1  A Lesson from Professor Marvel,
a.k.a. The Wizard of Oz . . . . . . . . . . . . . . . .5
*How to Customize Your Presentation*

2  Obama and You . . . . . . . . . . . . . . . . . . . . . .8
*The Most Persuasive Word*

3  The "So What?" Syndrome . . . . . . . . . . . .10
*...and How to Avoid It*

4  Beware of Jokes . . . . . . . . . . . . . . . . . . . . .12
*Dispelling a Common False Belief*

5  Presentation Advice from Abraham
Lincoln . . . . . . . . . . . . . . . . . . . . . . . . . . .14
*Clarity, Ownership, and Add Value*

6  It Ain't What You Say, It's How You
Say It . . . . . . . . . . . . . . . . . . . . . . . . . . . .16
*Lessons in Structure from Jeffrey Toobin and
Andrew Weil, M.D.*

7  Presentation Advice from Mark Twain . . . .18
*Brevity Takes Time*

8  Presentation Advice from
Mike Nichols . . . . . . . . . . . . . . . . . . . . . .20
*How to Find Value in Your Story*

9  Show versus Tell in Hollywood . . . . . . . . .22
*The Wrong and Right Way to Tell a Story*

10  Slogan Power .........................24
*Why the U.S. Army's "Be All That You
Can Be" Succeeded*

11  How Long Is Too Long? ...............26
*When in Doubt, Leave It Out*

12  The Elevator Pitch in One Sentence .....28
*How to Describe Your Business Succinctly*

13  Do You Know the Way to Spanish Bay? ...30
*The Correct Way to Practice*

14  Getting to "*Aha!*" .....................32
*The Magic Moment*

15  This Is Your Pilot Speaking ............34
*A Lesson in Flow from the Airlines*

16  Presentation Advice from the iPhone .....36
*Substance and Style in Your Story*

17  Presentation Advice from Steve Jobs .....38
*The Power of Positive Words*

18  Presentation Advice from Novelists I .....40
*Begin with the End in Mind, Then Write,
Rewrite, and Rewrite*

19  Presentation Advice from Novelists II ....42
*Storyboard and Verbalize*

20  Microsoft Slogans Score a Trifecta .......44
*Three Persuasive Techniques*

21  Presentation Advice from a Physician ....46
*Audience Advocacy*

22  Presentation Advice from a Politician ....48
*Audience Advocacy*

23  Ronald Reagan Meets Lenny Skutnik  . . . .50
     *The Catalyst of Human Interest Stories*

24  Human Interest Stories:
     A Double Advantage . . . . . . . . . . . . . . . . . .51
     *Two Ways to Use Anecdotes*

**SECTION II**       **Graphics: The Correct Way to Design
                      PowerPoint Slides**

25  The Presentation-as-Document
     Syndrome . . . . . . . . . . . . . . . . . . . . . . . . . . .55
     *Never the Twain Shall Meet*

26  Blame the Penmanship, Not the Pen . . . . .57
     *Operator versus Machine Error*

27  You Can't Use a Sentence As a Prompt! . . .59
     *Less Verbiage Is More Useful*

28  Baiting the Salesperson . . . . . . . . . . . . . . .60
     *Selling Is about In-Person Communication*

29  PowerPoint and Human Perception . . . . . .62
     *Scientific Support for Graphics Design*

30  PowerPoint Template:
     Combined Picture and Text . . . . . . . . . . . .64
     *The Best Positions for Pictures and Text*

31  Shady Characters . . . . . . . . . . . . . . . . . . . . .67
     *The Wrong Way and the Right Way
     to Build Text*

32  "I Can Read It Myself!" . . . . . . . . . . . . . . .69
     *Three Simple Steps to Avoid Reading
     Slides Verbatim*

33  A Case for Case I:
     Initial Caps or All Caps . . . . . . . . . . . . . . .71
     *Text Design in Presentations*

34  A Case for Case II: Serif or Sans  . . . . . . . .73
    *Font Design in Presentations*

35  What Color Is Your PowerPoint?  . . . . . . . .75
    *Contrast Counts*

36  Presentation Advice from Corona Beer  . . .78
    *Peripheral Vision Counts*

37  The Cable Crawlers . . . . . . . . . . . . . . . . . .80
    *How Television Animates Text*

38  Computer Animation  . . . . . . . . . . . . . . . .82
    *Three Simple Rules*

39  PowerPoint and the Military . . . . . . . . . . .84
    *Sometimes More Is More*

**SECTION III**        **Delivery Skills: Actions Speak Louder
                        Than Words**

40  The Art of Conversation  . . . . . . . . . . . . . .89
    *Eye Contact and Interaction Start
    at Infancy*

41  Presentation Advice from
    Edward R. Murrow . . . . . . . . . . . . . . . . . .91
    *The "Person-to-Person" Role Model*

42  Nonverbal Communication . . . . . . . . . . . .93
    *Look Them in the Eye* . . . . . . . . . . . . . . . . . .

43  Presentation Advice from Pianist
    Murray Perahia  . . . . . . . . . . . . . . . . . . . . .95
    *Concentration Creates Control*

44  Presentation Advice from Actress
    Tova Feldshuh . . . . . . . . . . . . . . . . . . . . .97
    *Concentration Creates Communication*

45  Presentation Advice from
    Michael Phelps and Dara Torres . . . . . . .99
        *How to Control Stress under Pressure*

46  Presentation Advice from
    Frank Sinatra . . . . . . . . . . . . . . . . . . . . .101
        *The Art of Phrasing*

47  Presentation Advice from Soprano
    Kiri Te Kanawa . . . . . . . . . . . . . . . . . . .103
        *The Importance of Breathing*

48  The One-Eyed Man . . . . . . . . . . . . . . . .105
        *Necessity Is the Mother of Invention*

49  Bill Clinton's Talking to Me! . . . . . . . . . .106
        *The Power of Group Dynamics*

50  Liddy Dole and Person-to-Person . . . . . .108
        *From Law School to the Republican
        National Convention*

51  Fast Talking . . . . . . . . . . . . . . . . . . . . . .109
        *Fun or Maddening*

52  Presentation Advice from Titian . . . . . . . .111
        *Position, Position, Position*

53  Presentation Advice from Musicians
    and Athletes . . . . . . . . . . . . . . . . . . . . . .113
        *The Value of Effortlessness*

54  Presentation Advice from Vin Scully . . . .115
        *From Reagan to Barber to Scully*

55  "Ya' Either Got It, or Ya' Ain't" . . . . . . . . .116
        *The Fear of Public Speaking Is Universal*

56  How to Eliminate the Fig Leaf . . . . . . . . .118
        *A Presentation Lesson from the Military*

57  Unwords . . . . . . . . . . . . . . . . . . . . . . . . . .120
    *Even Barack Obama Says Them*

58  To Slip or Not to Slip . . . . . . . . . . . . . . . .122
    *Been There, Done That*

59  The Free Throw  . . . . . . . . . . . . . . . . . . . .124
    *A Presentation Lesson from Basketball*

60  10 Tips for 30 Seconds  . . . . . . . . . . . . . .126
    *Help for Job Seekers*

61  You Are What You Eat . . . . . . . . . . . . . . . .127
    *10 Tips about Food and Drink
    in Presentations*

**SECTION IV      Q&A: Handling Tough Questions**

62  Speed Kills in Q&A . . . . . . . . . . . . . . . . . .131
    *The Vanishing Art of Listening*

63  A Lesson in Listening from
    Barack Obama . . . . . . . . . . . . . . . . . . . . . .133
    *How to Handle Multiple Questions*

64  If I Could Tell Jon Stewart... . . . . . . . . . . .135
    *Talk Shows Include Listening*

65  What Keeps You Up at Night?  . . . . . . . . .136
    *How to Handle the Most Frequently
    Asked Questions*

66  Spin versus Topspin . . . . . . . . . . . . . . . . .138
    *The Political World versus the Business World*

67  When Did You Stop Beating
    Your Wife? . . . . . . . . . . . . . . . . . . . . . . . . .140
    *How to Handle False Assumption Questions*

68  Madoff and Cramer Plead Guilty . . . . . . .142
    *How to Respond When Guilty as Charged*

69 Tell Me the Time, Not How to
Build a Clock . . . . . . . . . . . . . . . . . . . . .144
*Keep Your Answers Short*

70 Presentation Advice from Jerry Rice . . . .146
*Grasp the Question before You Answer*

71 Politicians and Spin . . . . . . . . . . . . . . . . .147
*Putting Lipstick on a Pig*

72 Murder Boards . . . . . . . . . . . . . . . . . . . .149
*How Elena Kagan Prepared for
Tough Questions*

73 Ms. Kagan Regrets . . . . . . . . . . . . . . . . .151
*Nonanswers to Tough Questions*

**SECTION V**      **Integration: Putting It All Together**

74 The Elephant . . . . . . . . . . . . . . . . . . . . . .155
*The Whole Is Greater Than the Sum
of the Parts*

75 Presentation Graphics Meet
Linguistics . . . . . . . . . . . . . . . . . . . . . . . .156
*Symmetry in Graphics Design*

76 One Presentation, Multiple
Audiences . . . . . . . . . . . . . . . . . . . . . . . .158
*12 Presenters, 12 Stories, 1 Set of Slides*

77 The Art and Science of
Oprah Winfrey . . . . . . . . . . . . . . . . . . . . .160
*The Secrets of Oprah Winfrey's Appeal*

78 Right or Left . . . . . . . . . . . . . . . . . . . . . .164
*The Deep Roots of Human Preferences*

79 Graphics Synchronization . . . . . . . . . . . . .168
*The Missing Link*

80  The House That Jack Built . . . . . . . . . . . . .170
        *Make All the Parts Fit*

    Footnotes . . . . . . . . . . . . . . . . . . . . . . . . . . .173

    Acknowledgments . . . . . . . . . . . . . . . . . . .177

    About the Author . . . . . . . . . . . . . . . . . . .178

    Index  . . . . . . . . . . . . . . . . . . . . . . . . . . . . .179

# Introduction

*I hear and I forget.*
*I see and I remember.*
*I do and I understand.*

Confucius
(551 B.C.–479 B.C.)

In the more than two decades I have been a presentations coach, many people have heard me tell them how to present effectively. They have also seen me show them, but the best results have come when I coached my clients to do—to put the techniques into action.

It worked. Those clients' presentations have raised hundreds of billions of dollars in public, private, and even not-for-profit financing; sold hundreds of thousands of products; formed thousands of partnerships; and gained approval for hundreds of internal projects.

Confucius was right about doing.

The techniques grew out of a variety of sources, going all the way back to my Master's studies in the Department of Speech and Drama at Stanford University and forward to my days as producer of public affairs programs at WCBS-TV in New York City. Although I didn't realize it at first, many of the techniques that go into producing a television program are the same as those required to create and deliver a winning presentation: telling a clear and concise story, designing effective graphics, presenting with confidence, and handling tough questions (the latter developed from my CBS assignment at the opposite end of the spectrum of *preparing* tough questions for the company's legendary inquisitor, Mike Wallace).

In 1988, I brought the accumulated techniques into the business world, where, after refining and applying them as a coach in the private

1

Power Presentations programs, I made them available to the public in three books: *Presenting to Win: The Art of Telling Your Story, The Power Presenter: Technique Style and Strategy,* and *In the Line of Fire: How to Handle Tough Questions.* Taken together, the three books span all the essential elements of any presentation.

An important adjunct to the techniques was to illustrate them with examples of other presentations as lessons of what to do and what not to do. Most of those examples came from my work with prior clients and with public figures. As the years progressed, I accumulated a substantial repertory of case studies from the business world and video clips from the political world. During that same time, I found additional examples in such diverse fields as current events, politics, science, art, music, literature, cinema, media, sports, and even the military. These variations from pure business cases proved to be even more valuable as coaching tools and lessons because they demonstrated the universal aspects of all human communication; in doing so they added significant dimension to the basic techniques.

This book consists of 80 new case studies from the front of the room. They are grouped into five sections mapped to the three original books, where you can find the basic techniques in full:

> Section I—Content: The Art of Telling Your Story
>
> Section II—Graphics: The Correct Way to Design Power-Point Slides
>
> Section III—Delivery Skills: Actions Speak Louder than Words
>
> Section IV—Q&A: How to Handle Tough Questions
>
> Section V—Integration: Putting It All Together

The first two sections relate to *Presenting to Win,* the third section, *The Power Presenter,* the fourth, *In the Line of Fire,* and the fifth incorporates all three books. I am confident that these diverse case studies will give you added depth and dimension for your presentation skills, as well as for all your communication skills. But for the techniques to work most effectively, you must also follow the advice of Confucius—and Nike—and just *do* it.

# Section I

## Content: The Art of Telling Your Story

# 1

## A Lesson from Professor Marvel, a.k.a. The Wizard of Oz

### *How to Customize Your Presentation*

In the opening scenes of *The Wizard of Oz* (F1.1), Dorothy runs away from her Kansas home and promptly encounters Professor Marvel, a seedy, itinerant con artist whose tacky traveling wagon advertises him as "Acclaimed by The Crown Heads of Europe." He offers his services to "Read Your Past, Present, and Future in His Crystal Ball."

Professor Marvel, played marvelously by Frank Morgan, takes one look at the naive girl, glances down at her suitcase, and says, "You're running away!"

Having missed his glance, Dorothy asks wondrously, "How did you guess?"

The Professor replies, "Now, why are you running away? No, no, don't tell me!" He looks off pensively, as if conjuring some magical power. Then, as if having divined a vision, he says conclusively, "They don't understand you at home!"

The wide-eyed girl smiles and says, "Why, it's just like you could read what was inside me!"

The Professor then offers Dorothy a crystal ball reading and asks her to close her eyes and concentrate. As she does, the Professor quickly rummages around in her basket. He then proceeds to describe what he pretends to see in the crystal ball, referencing the items in the basket.

Clearly, Professor Marvel is a charlatan, but we can learn a positive lesson from his trickery. He was able to connect with Dorothy and establish her trust by referencing relevant facts about her. The lesson here is that presenters can connect with their audiences by making references to relevant facts about individuals in the audience or about the audience as an affinity group.

Such connections are rare in today's presentations. Pressed by the demands of business, most presenters pirate their colleagues' slides, do minimal preparation, and then dump a load of generic data on their audiences, who, to all intents and purposes, would have been better off accessing a canned webinar.

Finding relevant facts that can customize any presentation doesn't require manipulative glances, the covert services of a private investigator, or an army of academic researchers. You can use seven simple techniques to build powerful connections with any audience.

1. ***Direct References.*** Schmooze. Just before your presentation, mingle with your audience. Chat with several different individuals. Talk with strangers and people you know. Ask them questions. Listen to their conversations. Gather information, names, and data points. Then when you step up to the front of the room, weave the names and information you've collected into your presentation.

2. ***Mutual References.*** Before your presentation, learn as much as you can about your audience. Visit their home pages. Cross-reference with a web search. Find links to persons, companies, or organizations that are in some way related to both you and your audience. Then at appropriate moments during your presentation, speak about those connections. Think of this as a tasteful, appropriate form of name-dropping.

3. ***Ask Questions.*** During your presentation, ask your audience questions; seek their opinions instead of answers to factual or true/false questions. Invite them to share their ideas, reactions, or stories.

4. ***Contemporize.*** On the day of your presentation, scour the Web, read the newspapers, listen to the radio, or watch television and find events or items that are relevant to your subject and your audience.

5. ***Localize.*** Prepare specific references to the venue of your presentation. Some information about a locale is common knowledge; some is available on the Web. In addition, you can go to the web site www.newseum.org/todaysfrontpages/flash/ (F1.2), where you can access that day's front pages of local daily newspapers around the country and around the world by city. Make your presentations fresh with up-to-the-minute references.

6. ***Data.*** Find specific information that links to and supports your message. The more closely linked your data is to your audience,

the better. If the information you cite is new to your audience, they will be impressed by the depth and currency of your knowledge. If your audience is already aware of the data, they will be pleased that you made the effort to relate to them.

7. ***Customized First Slide.*** Begin your presentation with a slide that includes the location, date, and logo of your audience or event.

You don't have to pose as a Professor Marvel, but you can make your audience marvel at your efforts to connect and personalize.

# 2

## Obama and You
### *The Most Persuasive Word*

In 2006, *Time* magazine picked "You" as the Person of the Year and published that issue with a Mylar patch on the cover that served as a mirror for readers to see themselves. (F2.1) The main theme of that issue was the personal power enabled by the World Wide Web, but it also drew attention to the power of the word *you*. If you search the Internet, you'll find tens of thousands of references to a Yale University study (unsubstantiated by Yale) ranking the 12 most persuasive words in the English language. *You* leads the list. (The others, in descending order, are *Easy, Money, Save, Love, New, Discovery, Results, Proven*, and *Guarantee*.) (F2.2) Unsubstantiated or not, President Barack Obama, an excellent speaker by any standard, fully understands and leverages the power of *you*.

Early in his campaign for the presidency, the *New Yorker* magazine ran a story about his campaign strategy:

> *Obama now tries to make a more personal connection with voters. In the past, he has been accused of making his campaign more about himself than about those who come to his rallies. Now the word* you *is mentioned as much as the word* I. *"You're not heard. They're not listening to what you need," he told a crowd assembled at a rodeo site in Fort Madison on a recent evening. "You deserve a president who is thinking about you."* (F2.3)

During his campaign, Mr. Obama's web site displayed a headline banner that read, "I'm asking you to believe. Not just in my ability to bring about real change in Washington ... I'm asking you to believe in yours."

Following his victory, Mr. Obama delivered his inaugural address, in which he used *you* or *your* 15 times, addressing the *you* of his audience, the "Americans of every race and region and station." He also used *us*, *our,* and *we,* pronouns that involve *you.* His speech included 23 instances of *us,* 62 instances of *we,* and a whopping 70 instances of *our,* which, when combined with the 15 instances of *you,* represent more than 7 percent of the total 2,388 words in his speech. (F2.4)

Although the historic day was his to celebrate, Mr. Obama kept his audience in mind. Flash-forward a year into his presidency. In a prime-time address to the nation from the United States Military Academy at West Point, Mr. Obama committed 30,000 more troops to fight the war in Afghanistan. The serious speech marked a significant change in his rhetorical style—from an audience-focused to a self-focused point of view. He used *you* only ten times, with four of them in the closing, "Thank you. God bless you. May God bless the United States of America. Thank you very much. Thank you." That left only six instances of the powerful word in the body of the speech. In sharp contrast, he used *I* 41 times.

The shift prompted Peggy Noonan, a *Wall Street Journal* political columnist—and former republican presidential speechwriter—to exclaim, "I, I—ay yi yi. This is a man badly in need of an I-ectomy." She went on to explain:

> George H.W. Bush famously took the word "I" out of his speeches—we called them "I-ectomies"—because of a horror of appearing to be calling attention to himself. Mr. Obama is plagued with no such fears. "When I took office ... I approved a long-standing request .... After consultations with our allies, ... I set a goal." That's all from one paragraph. He then used the word I in three paragraphs an impressive 15 times. "I believe I know," "I have signed," "I have read," "I have visited." (F2.5)

Granted, any senior executive—from the President of the United States to a midlevel manager in business—must take full responsibility for all decisions and actions, but Mr. Obama should not abandon the technique that helped get him elected.

The lesson for you is to involve your audiences by finding as many opportunities as you can to incorporate *you* in your presentations. In fact, adding *you* enhances any form of communication, from the soaring rhetoric of presidential addresses to the mundane routine of email. Try this simple practice: Before clicking Send on any message, make one final review to see where you can insert additional instances of *you.* Every time you do, you will heighten the impact of your message. Do the same throughout your presentations, and you will connect more often with your audiences and heighten your chances for success.

It is all about *you,* not hubris.

# 3 ――――――――――――――――――――――――――――――――

# The "So What?" Syndrome
## ... and How to Avoid It

How often have you sat in an audience listening to a presentation and said to yourself, "So what?"

If you are like most audience members—or like me, a presentations coach—your response to that question would likely range from "Quite often" to "Too often." Most presentations fail to address the audience's point of view. Worse still, most presentations, being all about the presenter, fail to offer benefits. Either failure produces the dreaded "So what?" Syndrome that produces disinterest and, ultimately, disconnect in the audience.

As a coach, I help my clients avoid failure by using "So what?" to a positive end. In coaching sessions with clients, I role-play an audience member: a potential customer for a product, a potential investor for a private or public financing, a manager for a project approval, a partner for a strategic alliance, or even a donor for a not-for-profit cause. In one session with a pharmaceutical company preparing for its Initial Public Offering (IPO), as the CEO rehearsed his road show, I assumed the role of an institutional investor. When the CEO described the clinical trials for his drug, he concluded his discussion by saying, "These trials prove that our drug is both safe and efficacious."

I stopped him and said, "So what?"

The CEO thought for a moment, and then added, "which will make our drug the preferred choice for physicians and generate significant revenues for our company." "Safe and efficacious" were valuable benefits for a physician who could prescribe the drug, but "generate significant revenues" is a benefit for an investor.

One of the most effective ways to avoid the "So what?" Syndrome in your presentation is to insert a WIIFY. WIIFY is an acronym of "What's in it for *you*?" and is based on the common axiom, "What's in it for *me*?"

The shift of the last word from *me* to *you* is deliberate because it shifts the focus from the presenter to the audience. The shift also leverages the power of *you*, the persuasive word you read about in the previous chapter. Think of the WIIFY as the ultimate benefit statement.

To insert a WIIFY in your presentation, pause your forward progress at a key point and start this sentence: "This is important to you because ...." Then finish it with a benefit to your target audience.

Or pause at another key point and pose this rhetorical question: "What does this mean to you?" Then answer it with a benefit to your target audience, a WIIFY.

Or pause at another key point and pose this rhetorical question: "Why am I telling you this?" Then answer it with a WIIFY.

Find as many key points as you can to insert WIIFYs in every presentation you give, and you will see the "So what?" interruptions vanish.

That last sentence is a WIIFY for *you*.

# 4

## Beware of Jokes
### *Dispelling a Common False Belief*

One of the most pervasive pieces of advice bandied about in the presentation trade is to start a speech or a presentation with a joke. Wrong! No one can guarantee the success or failure of any joke—certainly not a businessperson, but not even a professional comedian.

Consider Johnny Carson. The legendary talk-show host spent 30 years on late-night television telling jokes written by a crack team of professional, experienced comedy writers, but the jokes didn't always work. Fortunately, one of Johnny's greatest assets was his ability to recover from failed jokes. Whenever a scripted gag elicited no reaction or even groans from his audience, Mr. Carson mugged a silent take or made a comment about the bomb; either response often produced more laughter than some of the scripted jokes.

Consider one of Mr. Carson's most prominent successors, Jon Stewart of "The Daily Show." Although Mr. Stewart's adulatory studio audiences worship and roar at almost every word he utters, he occasionally produces a dud. Mr. Stewart recovers with one his many rubber-faced expressions or trademark cackles which, as with Mr. Carson, often produces more laughter than the planned gags.

If Johnny Carson and Jon Stewart can't guarantee a laugh, how can you?

Still, the temptation persists to break the ice in presentations with humor, to lighten up the proceedings, entertain, or engage the audience; all are noble intentions but still risky business. Even if a joke beats the odds and gets a laugh, the laughter is a digression from the main message of the speech or presentation.

The risk of humor is even riskier in today's globalized world and its diverse audiences. But diverse cultures still retain their original sensibilities, and comedy does not cross borders easily—even with a common language. U.S. humor and U.K. humour differ by much more than a single letter. If you have any doubt, watch Americans in the audience of a British music hall comedy. They are the only ones not laughing.

If, despite all these caveats, you still insist on telling a joke in your speech or presentation, make it self-deprecating. If you fail at making fun of yourself, your failure will be at your expense and not at your audience's.

# 5

# Presentation Advice from Abraham Lincoln

## Clarity, Ownership, and Add Value

Ted Sorensen, who was John F. Kennedy's special counsel and speechwriter, most notably for his historic inaugural address, studied the texts of Abraham Lincoln's speeches—at JFK's request—for style ideas. In an article for the *Smithsonian* magazine, Sorensen wrote that JFK "also asked me to read all previous twentieth-century inaugural addresses. I did not learn much from those speeches (except for FDR's first inaugural), but I learned a great deal from Lincoln's ten sentences." (F5.1)

The ten sentences Mr. Sorenson referenced are those of the memorable Gettysburg Address, but he also analyzed Mr. Lincoln's first inaugural address and, in particular, its last line. When Mr. Lincoln delivered the speech on March 4, 1861, the secession by Southern states was threatening to plunge the nation into Civil War. In his desire to conclude his oration on a note of unanimity, Lincoln carefully considered his wording.

"He needed no White House speechwriter," Mr. Sorensen said. "He wrote his major speeches out by hand, as he did his eloquent letters and other documents. Sometimes he read his draft speeches aloud to others, including members of his cabinet." In the case of the first inaugural address, Lincoln shared a draft with Secretary of State William Seward, who offered this suggestion for the closing sentence:

*The mystic chords, which, proceeding from so many battlefields and so many patriot graves, pass through all the hearts and all the hearths in this broad continent of ours, will yet harmonize in their ancient music when breathed upon by the guardian angels of the nation.*

Mr. Sorensen added that "Lincoln graciously took and read Seward's suggested ending, but, with the magic of his own pen, turned it into his moving appeal:"

*The mystic chords of memory, stretching from every battlefield and patriot grave to every living heart and hearthstone all over this broad land, will yet swell the chorus of the Union, when again touched, as surely they will be, by the better angels of our nature.*

Presenters can learn a triple lesson:

1. **Clarity.** Mr. Seward's version of the sentence rambled on in a long, convoluted series of subordinate clauses. In Mr. Lincoln's rewrite, each clause builds upon the other in a straight ascending arc to culminate on a strong, affirmative note. The lesson for you is not about poetic or lyrical progression, but about succinctness. All too often, presenters ramble on in long convoluted sentences that wend their way into the weeds. Speak in crisp, clear, and brief phrases. Make your points and move on.

2. **Ownership.** Presenters must take responsibility for their own presentations. You cannot and must not attempt to deliver someone else's presentation, nor can you delegate the creation of your presentation to another person or group. This does not mean that you should write out every word by hand, as Mr. Lincoln did, but do participate in the development of the content. However, you *must* practice your presentation as Mr. Lincoln did. Speak it aloud—to others or to yourself—several times in advance, because only then will you feel comfortable and be effective when you deliver it.

   I realize that this recommendation presents a challenge in the corporate world, where, in the interest of unified messaging, presentations are usually generated by a central marketing group and then distributed throughout the organization. This not to say that you should depart from the company story, but to take that story, tweak it to your style and practice it aloud.

3. **Add Value.** Mr. Sorensen researched diligently on behalf of JFK, and so did Mr. Lincoln on his own behalf. Mr. Sorensen noted that Mr. Lincoln had a "willingness to dig out facts (as his own researcher)." No one knows your own subject as well as you do, but don't rely only on your knowledge alone; find additional supporting information to add value to your story.

If Abraham Lincoln could do all this and run the country, you can find the time to take charge of your presentation.

# 6

## It Ain't What You Say, It's How You Say It

### *Lessons in Structure from Jeffrey Toobin and Andrew Weil, M.D.*

The first commandment in all communications is that the messenger is just as important as the message—or, in the vernacular, it ain't what you say, it's how you say it.

Jeffrey Toobin and Andrew Weil, M.D., are, by any standard, on the A-list of public speakers. Each man has what is known on the keynote speaking circuit as a solid platform. The term refers to a large installed base of loyal followers built by frequent access to and exposure in the media. Mr. Toobin, a legal political analyst for CNN, appears regularly on that cable channel; and Dr. Weil, a trusted health advisor (as his trademark reads), runs a vast online marketplace that sells personal care products, vitamins, and cookware. Each man also has a string of best-selling books, including Mr. Toobin's *The Nine* and Dr. Weil's *Natural Health, Natural Medicine.* All these factors create an attractive draw for their public speaking engagements.

As part of the San Francisco Bay Area's lecture series, both men drew sold-out crowds of more than 3,000 people to the historic Frank Lloyd Wright–designed Marin Civic Center. Both men are dynamic speakers, but their individual styles present an interesting study in contrasts.

Of the two, Mr. Toobin was more effective, despite that fact that Dr. Weil's subject—health—was of greater intrinsic and personal value to the audience than Mr. Toobin's drier subject—the Supreme Court. The difference was their organizational structures.

During his hour, Dr. Weil touched on a wide array of subjects, including vitamins, diet, health care reimbursement, and even an audience-participation breathing exercise that had all 3,000 people huffing and puffing along with him. Although Dr. Weil covered each topic thoroughly, each one stood alone, without a link or transition to the next, making it challenging to follow. Mr. Toobin, on the other hand, had a single theme: the composition of the Supreme Court. Although he moved backward and forward in time, discussing the varying combinations of the nine justices in different decades, and although he peppered each story with human interest anecdotes, his speech was easy to follow because each move and anecdote supported and pivoted around the central theme. Mr. Toobin held the audience in rapt attention throughout his speech.

What made Mr. Toobin's structure work? How could Dr. Weil have improved his?

Speakers can select one of 16 different Flow Structures, or logical templates, to organize the diverse components of any story into a clear roadmap for audiences. *Presenting to Win* describes all 16 in detail, but 2 of the simplest and most frequently used are the Numerical and Chronological Flow Structures. David Letterman uses the Numerical Flow Structure in his nightly "Top Ten," and Stephen Covey uses it in his *Seven Habits of Highly Effective People*. The Chronological Flow Structure is often used in business to describe a company's track record, present position, and future direction, or to trace a product evolution from inception to release, to upgrade.

Jeffrey Toobin chose Numerical as his primary Flow Structure with Chronological as his secondary. His central theme was the composition of the nine justices along liberal and conservative lines. The shifting balance of power among the nine is a constant source of dramatic tension that drives presidential elections, political parties, and many impassioned contending constituencies. Mr. Toobin discussed the various majorities and minorities among the justices at different points in time. Because of his central focus on the total number, he was able to jump backward and forward among different decades and even add sidebar human interest stories, yet still maintain a clear narrative thread.

Dr. Weil could have created continuity by emulating David Letterman and Jeffrey Toobin by choosing Numerical: assigning a number to the diverse health topics he discussed, with "Six (or Seven) Health Challenges." Then if he were to "Tell them what he was going to tell them" at the beginning, countdown as he "Told them," and then "Tell them what he told them" in summary, his audience would have followed along easily. Or, as Aristotle advised 2,300 years ago, Dr. Weil would have created a clear beginning, middle, and end.

Aristotle is considered as a classic because his wisdom endures.

# 7

## Presentation Advice from Mark Twain

### *Brevity Takes Time*

The celebrated American author Mark Twain was also a most prolific writer. Amazon lists more than 12,000 books consisting of various editions of Mr. Twain's own works and works about him. So great was his output that his quotes alone—some actual, some apocryphal—have even more references on the Internet than do his books.

One of his most famous quotes, which is quite applicable to presentations, came from an exchange Mr. Twain had with his publisher. The publisher sent the author a telegram reading:

*NEED 2-PAGE SHORT STORY [IN] TWO DAYS.*

The writer sent back a telegram reading:

*NO CAN DO 2 PAGES TWO DAYS. CAN DO 30 PAGES 2 DAYS. NEED 30 DAYS TO DO 2 PAGES. (F7.1)*

Mr. Twain's pithy nineteenth-century observation captures the essence—and the chronic problem—of twenty-first-century business communications. Although email and Twitter have instilled a drastic decline in the verbiage (and the style, spelling, punctuation, and courtesy—but those are subjects for another time) of today's exchanges, the most mission-critical of all business communications, the presentation, still suffers from Mr. Twain's dilemma. The pressures and pace of modern life allow very little time to prepare pitches. As a result, the quick-and-dirty approach inevitably produces sagas that approach the length of doctoral dissertations, the equivalent of delivering a treatise on how to build a clock when all that is needed is to tell the time.

We can measure the consequence of this dilemma in another manufacturing operation, that of automobile wheels: The longer the spoke, the bigger the tire.

Today's business audiences, driven by their own daily pressures, do not have the time—or the patience—to listen to the entire history of Western civilization when you take the floor.

Solve Mark Twain's dilemma for your presentations. Invest the time and effort to prepare for your mission-critical pitch. Start early and do several drafts. Don't leave the preparation time for your presentation until the flight to the city in which you will be delivering it. That approach will produce an epic of encyclopedic size—and a reaction of yawning sighs.

Oh, I know, your plate is very full, but which of your many daily tasks has as much impact as the brief window of opportunity you have when you present to decision makers? Andy Warhol's much-referenced 15 minutes of fame have their equivalent in the precious moments you have in front of your live audience. Make those moments count by preparing thoroughly.

It will be well worth your while—and, even more important, your audience's while.

# 8

# Presentation Advice from Mike Nichols
## *How to Find Value in Your Story*

Mike Nichols, the noted director of numerous Hollywood films (including *The Graduate, Catch-22,* and *Who's Afraid of Virginia Woolf?*), Broadway comedies, and television productions, is a master of his craft, with many Oscar, Tony, and Emmy awards to his credit. The creative approach Mr. Nichols uses to develop his theatrical stories provides an object lesson to help you develop your presentation story.

In an interview with the *New York Times*, Mr. Nichols described how he prepares for a film: "I really do think it's important to sit with a text for as long as you can afford to, reading and talking." He called this process "naming things," which he described as "just explaining what happens in every scene." (F8.1)

You can use the "naming things" process in preparation for your presentation, but do so *after* you have shaped your story. Mike Nichols employs his process with a shooting script in hand. In that same manner, you can use his approach only when you have evolved your presentation to an equivalent stage by going through these important developmental steps:

- Set the context, the presentation objective, and how it relates to your audience.
- Brainstorm all the potential ideas that support your objective and provide benefits to your audience.
- Distill the essential ideas and discard the excess.
- Structure those final ideas into a logical flow.
- Design graphics that illustrate your story.

When you have accomplished this, you can proceed to implement your own "naming things" process. Look at each slide in your deck and decide its main point. Then go back through the deck and speak your narrative aloud in rehearsal, stating those main points. As you move through the deck, maintain the flow by making each slide relate to the preceding and following slides. Then go back through the deck once more and, this time, punctuate each slide with either a restatement of your objective or a benefit to your audience. This puts the icing on the cake and lifts your presentation to its optimal level.

Contrast this comprehensive approach with the more typical method of a last-minute cobbling together a disparate assortment of begged, borrowed, or stolen slides, and then standing up in front of a mission-critical audience and reading the slides to them verbatim.

Although an expert at comedy, Mike Nichols would not be amused.

# 9

## Show versus Tell in Hollywood
### *The Wrong and the Right Way to Tell a Story*

Lesson One in Screenwriting 101: *Show Don't Tell*. In a well-made film, the story advances by action. In a lesser film, the story advances by exposition, with the characters describing the action. In an inferior film, the story is advanced by an unseen narrator.

The latter technique is drawn from books, where, because of the absence of visual images, the unseen author must describe the images and the action. In books, the art of telling the story is in the author's narrative word craft; in films, the art of telling the story is in the director's camera and editing choices. In presentations, the art is in the value the presenter adds beyond what the audience sees on the slides.

Three major films provide three different directorial approaches to cinema storytelling: Gus Van Sant's *Milk*, David Fincher's *The Curious Case of Benjamin Button*, and Woody Allen's *Vicky Cristina Barcelona*. The first two use a principal character as an onscreen narrator to advance the story. In *Milk*, Sean Penn carries the narrative as the title character dictating his story into a tape recorder; in *Benjamin Button*, Cate Blanchett's character carries the narrative as an old woman revisiting her life through her scrapbook. These films are, by any standard, excellent productions. Both have powerful performances, rich production values, and important themes, but both repeatedly interrupt the forward progress by returning to the narrator to tell the story. Each film has enough going on in the action to propel the story forward without having to resort to this disruptive narrative device.

The Woody Allen film also has a narrator, but it is an off-screen male who comments on the story instead of telling it. Mick LaSalle, the *San Francisco Chronicle* movie critic, noted this in his review: "Voice-over narration gets a bad rap because it's often added as an afterthought to films that don't hang together in the editing. But in *Vicky Cristina*

*Barcelona,* the narration was built into the design, and it's used extensively and effectively, placing us securely in the story." (F9.1)

Follow Woody Allen's example: Add value to your story by expanding beyond what is on your slides. Show your story in action by providing examples, case studies, analogies, analysis, benefits, and conclusions.

Show, don't tell.

# 10

## Slogan Power

### *Why the U.S. Army's "Be All That You Can Be" Succeeded*

The U.S. Army's "Be All That You Can Be" slogan, which ran for more than 20 years—until it was replaced by the far less memorable successor, "Army of One,"—became as well known as any commercial brand.

Bill Peacock was the Assistant Secretary of the Army for Manpower and Reserve Affairs who helped spearhead the "Be All That You Can Be" campaign. He provided a back story that offers a lesson for any communication medium.

In the late 1970s, the Army was experiencing serious defections from its ranks and was having trouble attracting replacements. Almost half of those ranks were populated by high school drop-outs, a demographic most likely to drop out of the Army as well. The churn was costing the government almost $2 billion a year. The other armed services—the Air Force, the Marines, and the Navy—were experiencing much less churn because their ranks were filled with higher percentages of the more stable population of high school graduates. Under Mr. Peacock's leadership, the Army decided to create an enlistment campaign to attract more of that demographic group. They considered many aspects of importance to the target group of 18-year-olds and eventually focused on three central themes:

- Patriotism
- Manliness
- Personal aspiration

The Army then conducted market research to see which of the three themes resonated with the target audience. They tested them in surveys at hundreds of shopping malls around the country, and the third won, hands down. The personal aspiration theme became the "Be All That You Can Be" slogan. It was expressed in a full multimedia suite, using real soldiers instead of actors, complete with catchy theme music and professionally produced video spots.

The Army then engaged the N.W. Ayer advertising agency to roll out the campaign in many media outlets, mostly television. Within about six months after the launch, the percentage of high school graduates in the Army rose from less than 50 percent to more than 70 percent—an astonishing success.

An important by-product of the campaign was improved morale within the ranks, as evidenced by a significant uptick in reenlistment rates.

We can learn a double lesson in this story: The "Be All That You Can Be" theme succeeded because it was aimed directly at its target audience—the young men who aspired to improve their position in life. The slogan was essentially a benefit statement or WIIFY, the subject you read about in Chapter 3, "The 'So What?' Syndrome"; it also incorporated the powerful word *you*.

Whenever you present, offer your audience a WIIFY or multiple WIIFYs, and say *you* frequently.

Then *you* can be all that *you* can be.

# 11

## How Long Is Too Long?
### *When in Doubt, Leave it Out*

In 1988, Bill Clinton, then the governor of Arkansas, gave a nominating speech for Michael Dukakis at the Democratic National Convention. By convention rules, Mr. Clinton was allowed 15 minutes, but he brought an 18-page speech. According to the *New York Times* (F11.1), he read almost every word, rambling on for so long that the delegates began to chant, "We want Mike!" Ignoring their chants, Mr. Clinton went on and on and on. When, after 30 minutes, he finally said, "In conclusion ..." the crowd roared their approval.

Years later, in his autobiography, Mr. Clinton admitted, "It was 32 minutes of total disaster." (F11.2)

For his first venture onto the national political stage, Bill Clinton neglected to take a lesson in brevity from other national leaders' speeches:

- Abraham Lincoln's Gettysburg Address: 272 words
- Abraham Lincoln's second inaugural address: 700 words
- Winston Churchill's "blood, toil, tears, and sweat" speech that launched Britain's entry into World War II: 627 words
- John F. Kennedy's inaugural address: 13.5 minutes

Barack Obama, Mr. Clinton's Democratic successor, learned well from his predecessors. His now-famous speech at the 2004 Democratic National Convention ran 16.5 minutes, and his own now-historic 2009 inaugural address ran 18.5 minutes.

Bill Clinton ultimately learned his lesson. As president, he became a charismatic speaker, able to captivate any audience. After he left office, he leveraged that skill on the speaking circuit, with fees for his keynotes earning him eight figures annually.

The best business example of the value of brevity comes from that most mission-critical of all presentations—the IPO road show. I have had the privilege to work with many companies developing their road shows. When a company offers shares of its stock to the public, the CEO and the CFO go on an arduous two-week tour, delivering 70 or 80 iterations of the same presentation. Inevitably, as the tour proceeds, each iteration of the presentation gets longer and longer in a phenomenon known as "presentation creep." Targeted at 20–25 minutes, by the end of the road show, some presentations have been known to run as long as 40 minutes. Except for one company.

Let's call them XYZ. During XYZ's road show, I received an excited call from an investment banker friend in New York who had just attended the luncheon presentation. He said, "This was the best road show you've ever coached!"

"Why?" I asked.

"Because they did it in 15 minutes, and the investors were delighted to get back to their offices."

It is doubtful that you will face the same challenges and stakes in your presentations as national leaders and IPO road show executives, but you can learn from the masters. Your audience doesn't need to hear every encyclopedic detail of your business proposal. They have neither the time nor the patience—especially in this Twitter-driven world—to listen to long presentations.

Keep brevity in mind while you are developing your presentations, and when you are finished and ready to present, take one more look at it and find more material you can omit. Follow Ludwig Mies van der Rohe's famous advice *Less Is More*, and its corollary: "When in doubt, leave it out."

# 12

## The Elevator Pitch in One Sentence
### *How to Describe Your Business Succinctly*

In a *Wall Street Journal* article about the Herculean tasks facing President Barack Obama—the economic crisis, the environment, health care reform, Iraq, Afghanistan, Iran, and North Korea—columnist Peggy Noonan referenced Clare Boothe Luce, a noted twentieth-century playwright, journalist, ambassador, and congresswoman. Ms. Luce "told about a conversation she had in 1962 in the White House with her old friend John F. Kennedy. She told him, she said, that 'a great man is one sentence.'"

Ms. Noonan defined that one sentence as "leadership [that] can be so well summed up in a single sentence that you don't have to hear his name to know who's being talked about. 'He preserved the union and freed the slaves,' or, 'He lifted us out of a great depression and helped to win a World War.' You didn't have to be told 'Lincoln' or 'FDR.'" (F12.1)

As a conservative columnist and former speechwriter for Ronald Reagan and George H. W. Bush, Ms. Noonan said that Mr. Obama was replacing "his sentence with 10 paragraphs" and recommended that he should try to narrow his focus because "an administration about everything is an administration about nothing." Jon Stewart, a longtime supporter of Mr. Obama, echoed Ms. Noonan on an episode of his *The Daily Show*, saying, "So, Mister President, while I am impressed with your Renaissance man level of knowledge in a plethora of subjects, may I humbly say, 'That's great! Now fix the economy!'" (F12.2)

The one-sentence recommendation is also applicable to business, with particular regard to the universally referenced elevator pitch—so named to refer to the way you'd describe your business if you stepped into an elevator and suddenly saw that hot prospect you've been trying to buttonhole. The allusion is intended to limit the pitch to the length of an elevator ride. Unfortunately, most such pitches in business are often as long as elevator rides in the Empire State Building, the equivalent of what Ms. Noonan calls President Obama's "10 paragraphs."

A guide to help you create a succinct elevator pitch can be found in the words of Rosser Reeves, a prominent advertising executive with the Ted Bates agency during the middle of the twentieth century.

Mr. Reeves coined the term "Unique Selling Proposition (USP)," defined in his biographical sketch in *Advertising Age* magazine as "the one reason the product needed to be bought or was better than its competitors." (F12.3)

These USPs often took the form of slogans. Reeves oversaw the introduction of dozens of slogans, some that exist to this day, such as M&M's® "Melt in your mouth, not in your hand." He argued that advertising campaigns should be unchanging, with a single slogan for each product.

To pitch or describe your own business, develop your own USP along Rosser Reeves's guidelines. One of the most common complaints about presentations is, "I listened to their pitch for 30 minutes, and I *still* don't know what they do!"

The USP is what they do.

The one sentence Peggy Noonan recommended for Barack Obama would undoubtedly satisfy Jon Stewart—as well as every man and woman in America: "He brought America back from economic collapse and kept us strong and secure in the age of terror."

# 13

## Do You Know the Way to Spanish Bay?
### *The Correct Way to Practice*

The Inn at Spanish Bay in Pebble Beach, California, is 85 miles from the heart of Silicon Valley. This proximity, along with its first-class golf course and attractive seaside location, makes the resort a popular destination for conferences run by the Valley's many technology organizations. As a presentations coach, I am often invited to such conferences to give a presentation about how to give a presentation. I usually deliver the same subject matter, adapted for and oriented to each unique audience. My subject matter is drawn from material I have been delivering for more than two decades, and so I am quite familiar with the content.

Nevertheless, I practice each of these presentations thoroughly, utilizing a technique I recommend to every client I have ever coached and I now recommend to you: Verbalization, which means speaking your presentation aloud in advance just as you will to your actual audience, and doing it many times. This powerful rehearsal methodology has analogues in sports, music, theater, and adages: Practice makes perfect.

Unfortunately, the way most businesspeople rehearse their presentations is by clicking through the slides and saying something like, "Okay, with this slide I'm going to say something about our sales revenues ... and then with this slide, I'll say something about our path to profitability ... and then with this next slide, I'll show a picture of our lab and talk a little about R&D."

Sound familiar? As a form of rehearsal, it is completely unproductive. Talking about your presentation is not an effective practice method for presenting, any more than talking about tennis would be a good way to improve your backhand.

An even more common presentation practice is mumbling. The presenter clicks through the slides on the computer or flips through the pages of a hard copy of the slides while muttering unintelligible words.

Neither of these methods is Verbalization. Both are counterproductive because they reinforce negative behavior.

Jason Trujillo of Intel Corporation described this behavior as "practice makes *permanent*," a variation of the 2,000-year-old words of Publius Syrus, "Practice makes perfect." If you mumble, you reinforce mumbling. If you Verbalize your words just as you will say them in front of an actual audience, you reinforce the correct words.

As a demonstration of the power of Verbalization, let's turn to a literary device used by Robert Greene in his bestselling book *The 48 Laws of Power*. Mr. Greene describes each of the laws and then proves the power of the law by illustrating the consequences of what he calls "transgression of the law." (F13.1) Here's what happened when I transgressed my own advice and failed to practice my own presentation aloud in advance:

Of the many presentations I have delivered at the Inn at Spanish Bay, four were to the annual conferences of one major investment bank. Each of those four conferences had the same agenda, so each of my presentations covered relatively the same content. True to my own advice, I practiced my presentations aloud and did so during the nearly two-hour drive from my Silicon Valley office to the resort. (The advent of Bluetooth made speaking alone in my car appear less strange to other drivers on the road.) I Verbalized each of those annual presentations—except for the fourth.

Just as I was about to leave for that event, an important business matter arose and I had to spend most of the drive time on my mobile phone dealing with the matter instead of Verbalizing my presentation. Having transgressed, I paid the price. When I got to the hotel and stepped up to the dais to speak, my delivery was choppy. Imagine that: My delivery was choppy with familiar content, and I am an experienced presentations coach who presents almost every business day of his life.

Do as I say, not as I did on the way to Spanish Bay: Verbalize for every one of your presentations.

# 14

## Getting to "Aha!"
### *The Magic Moment*

Dr. John Kounios, a psychologist at Drexel University in Philadelphia, defined the "Aha!" Moment in a *Wall Street Journal* article as "any sudden comprehension that allows you to see something in a different light .... It could be the solution to a problem; it could be getting a joke; or suddenly recognizing a face." (F14.1)

*The "Aha!" Moment*

According to William Safire, the master etymologist of the *New York Times,* the first person to express the moment was Chaucer in a fox hunt in his *Canterbury Tales.* (F14.2) Archimedes undoubtedly experienced it when he noticed the displacement of water in his bath, as did Sir Isaac Newton when he saw an apple fall from a tree, and Alexander Graham Bell when he called to his assistant, "Watson, come here; I want you."

*The "Aha!" Moment*

Oprah Winfrey and Mutual of Omaha engaged in a protracted legal battle over the advertising use of the phrase until they finally settled out of court.

*The "Aha!" Moment*

History is filled with quests for the defining moment—that magic instant of Eureka or epiphany, that sudden turning or tipping point.

You can achieve the presentation equivalent of the "Aha!" Moment with your audiences if you learn one very simple technique—but be forewarned that the technique breaks rank with common practice in business today. Standard Operating Procedure is to load up the PowerPoint deck with all the information that the presenter thinks an audience needs to evaluate a proposal and make a decision. In other words, the slide show is meant to stand alone. You will read about the folly of this approach in the next section on graphics, but it will be from the design point of view.

For now, let's look at the presenter's point of view. If the slide show is correctly viewed solely as support for the presenter—as it should—then the presenter's narrative can go beyond the information on the slides to provide substance and even add value. The presenter can go even further to lead the audience to a conclusion about the slide with information that does not appear on the slide.

For example: The CEO of a start-up company seeking financing makes a pitch to a venture capital company. During the presentation, the CEO puts up the simple slide in Figure 14.1:

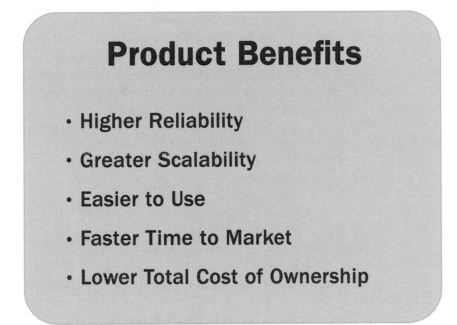

**Figure 14.1  Presentation slide**

The CEO discusses each of the benefits briefly and then summarizes, "So you can see that our product provides a rich set of benefits to our customers."

All very well and good, but imagine if the presenter were to add the following statement in the narrative only: "These benefits bring our company repeat business, repeat business brings us recurring revenues, and recurring revenues grow shareholder value."

*The "Aha!" Moment.*

# 15

## This Is Your Pilot Speaking
### *A Lesson in Flow from the Airlines*

Picture this: You've settled into your airline seat, buckled your seat belt, and turned off your laptop computer, your mobile device, and your iPod. With nothing else to do, you start to read today's newspaper. The airliner pulls away from the gate, taxis onto the runway, and rolls into a long line behind other jets waiting to take off. And it waits. And it inches forward. And waits ....

As it waits, the jet's giant motors keep running at idle, their vibrations gently rocking the cabin with a constant low hum. And you continue reading. And the motors continue to idle in their repetitious drone. And you drift off into the sweetest, deepest, and most delicious sleep you've had since you were in the cradle.

"Good afternoon, ladies and gentlemen, this is your pilot speaking," a deep, disembodied voice squawks from the tinny speaker directly over your head. You try not to awaken, but the cheery voice continues, "We're sorry for the delay, folks, but we've now reached our cruising altitude and we're going to do our darndest to make this big bird fly real fast and get you to Chicago on time."

The voice pauses, and you drop back into your reverie.

But, a moment later, a voice returns: "Our flight route today will take us up over the Carquinez Strait and on up to Boise where we'll make a right turn and head straight on over to Sioux Falls, then Casper, Dubuque, and right on into Chicago. Our flying time will be ...."

"Carquinez ... Boise ... Sioux Falls ... Casper ... Dubuque ...." Who cares?

Why do pilots do this? Every one of them does it, every time. It's as predictable as the flight attendants' canned speech about oxygen masks.

I have a theory about this routine. I have no proof for my theory. Nor do I have any knowledge of any internal airline policy that requires their pilots to go through this interruptive ritual. Nor do I seek proof of my theory. I am perfectly content to accept this rationale:

Somewhere deep in the depths (or up in the highest reaches) of the corporate infrastructure of the airlines or the Federal Aviation Administration, someone or some group—marketing, public relations, human resources, legal, or all of the above—decided that having the pilots announce the flight route would reassure the passengers that the airline is in complete control.

So it is not just your pilot speaking; it is the pilot speaking on behalf of the entire corporate and governmental infrastructure.

Presentation audiences are very much like airline passengers. Both groups are stuck in their seats for the duration of a fixed period of time, and both are at the mercy of another party to lead the way. Clearly, the airlines, their intrusive announcements notwithstanding, are merciful. Presenters are not.

Picture this: You've settled into your audience seat, and turned off your computer and your smart phone. With nothing else to do, you start to listen to the presenter. After a moment or two, you say to yourself, "Wait a minute! What does *that* have to do with the previous slide? How did you get here from there?"

Clearly, the presentation has no logical route. Every slide makes a point completely different from the preceding slide, and the presenter has made no effort whatsoever to link the two slides. In fact, the presenter begins each slide with the same meaningless phrase: "Now I'd like to talk about ...."

This is not a problem for presenters because they know the interstitial connections between the slides. It is a major problem for the audience because each slide begins the presentation anew, forcing the audience to work hard to find the logic. If you make it hard for your audience, they will make it hard for you.

This disconnectedness in presentations is as universal as route announcements are in air travel. A very simple way to create a clear flight path for your presentation is to bundle all the individual parts of your story within an overarching structure. Make the relationships among all the parts crystal clear. Encompass them in a roadmap.

Become the pilot for your audience. Announce your route and track them through it step by step. As you move from slide to slide and section to section, use transitional statements that make the relationships between them and the overall flow clear.

Make it easy for your audience to follow, and they will make it easy for you. Keep them from slipping off into dreamland. Leave the sleeping to airline travel and the driving to *you*.

# 16

## Presentation Advice from the iPhone
### *Substance and Style in Your Story*

The colossal success of the iPhone and its big brother, the iPad, are due in large part to their high-concept design, which, in turn, offers a lesson in crafting your story. This analogous viewpoint comes from Bill Portelli, the president and CEO of CollabNet, a company that provides a platform for distributed software development. Mr. Portelli noted that the iPhone has three basic qualities:

- Simple
- Intuitive
- Assumes the intelligence of its end user

Anyone who owns an iPhone—or who has looked over a shoulder at someone else's iPhone—will agree that Mr. Portelli's observation is spot on. A well-told story should have the same basic qualities:

- **Simple.** Most presenters make the fatal mistake of overwhelming their audiences with too much information, also known as TMI. In the parlance of the highest of high technology, you have to make your story easy enough for your mother to understand.

- **Intuitive.** Your story is brand new to your audience, and so they need a logical context to process your content. Provide them with a clear flow, and keep them in the flow by using linkages as you move through the individual parts of your story. Make it easy for them to follow, and they will follow your lead. Make it hard for them, and they will make it hard for you.

- **Assume Intelligence.** Give your audience more than the usual boilerplate features, benefits, and facts. Your audience has been there, done that, and they get it. They need and can process more. Add value and dimension to your story with examples, analogies, anecdotes, evidence, current data, and customized content.

Apple touts the iPad's multitouch screen as one of its main features, which provides another analogous piece of advice to presenters: Connect with your audience at multiple touch points—through their eyes, their ears, and their brains.

When you tell your story, be a Mac.

# 17

## Presentation Advice from Steve Jobs
### *The Power of Positive Words*

One of the bestselling books in the "Running Meetings and Presentations" category on Amazon is *The Presentation Secrets of Steve Jobs: How to Be Insanely Great in Front of Any Audience*. Author Carmine Gallo offers his readers presentation lessons from the charismatic Apple CEO. Another lesson comes from Neil Curtis, a conceptual artist who created a mashup of the launch presentation for the iPad. The clever video clip (available on YouTube) strings together the scores of adjectives that Mr. Jobs and his Apple copresenters used during the event—multiple iterations of *extraordinary, phenomenal, great, awesome, super, amazing,* and *terrific.* (F17.1) All are quite appropriate for a product launch.

As Miguel Cervantes said, faint heart ne'er won fair lady.

But too much of a good thing can become a bad thing. Too much milk or too many carrots can make you sick. Self-praise is no compliment. Furthermore, after the Internet bubble, the real estate bubble, and Bernie Madoff's Ponzi scheme, the public—particularly the technology public—is very wary of hyperbole. *Vaporware* is a dirty word.

Such cautionary advice might seem a contradiction in terms coming from someone whose company's first name is "Power," but I work both sides of the aisle, providing advice on how to add punch to presentations as well as to tone down any oversell. Given my stock in trade of IPO road shows, the latter is an absolute requirement. After the painful burst of the Internet bubble and the rigorous remediation of the Sarbanes–Oxley Act, no company can put itself at risk with even a hint of hyperbole. As a coach, I closely monitor that factor.

However, most of my coaching is focused on the other side of the aisle: adding punch. Most of that effort is in elevating the level of assertion in a presenter's narrative language; not to insert Jobs-like adjectives.

This is an equal opportunity task. All presentations, from the highest-profile IPO road show to the most basic product pitch, tend to use weak verbiage that diminishes the strength of the message.

Consider the three most commonly used phrases in presentations:

- "We believe ..."
- "We think ..."
- "We feel ..."

The use of these phrases increased dramatically in this Sarbanes–Oxley era to avoid forward-looking statements, but the phrases have been with us far longer. I know not from where they came, but their presence is pervasive. They are all formations in the conditional tense that impart doubt to any words that follow. Power Presentations are not doubtful presentations.

This is not to recommend that presenters should switch to the declarative tense and start forecasting results or selling snake oil, but to switch to more assertive words within the conditional tense:

- "We're confident ..."
- "We're convinced ..."
- "We expect ..."

Now *that's* a Power Presentation.

# 18

## Presentation Advice from Novelists I

### *Begin with the End in Mind, Then Write, Rewrite, and Rewrite*

An article in the *Wall Street Journal* described the creative processes used by several different novelists. As you might expect, their methods were as varied as their literary styles, ranging from preferred writing materials to favorite venues, and even to the most productive times of day. But all the writers shared one common technique: They compose many drafts of their work. (F18.1)

One of the foremost proponents of rewriting is John Irving, the author of the bestselling novel *The World According to Garp* and 14 other novels. Mr. Irving states his method for all the world to see on his web site: "Rewriting is what I do best as a writer. I spend more time revising a novel or a screenplay than I take to write the first draft." (F18.2)

Granted, novelists have the luxury of time that few businesspeople do, but rewriting is just as important in business because writing presentations is also a creative process. An indispensable part of the creative process is spaced learning, or the practice of pausing between drafts to enable the ideas to ripen. The opposite of spaced learning is cramming. Every professional writer, from novelists to journalists, understands and practices spaced learning; unfortunately, businesspeople, driven by the rapid pace of business, do not.

How can you apply spaced learning in business? Start with email. I have an ironclad rule about my electronic correspondence. For any email of substance: an after composing my message, I save it as a draft for an interval of anywhere from a few minutes to a few hours. Inevitably, whenever I return, I find words, phrases, sentences, or expressions that need a rewrite.

If spaced learning is helpful in the short form of email, imagine how much more helpful it can be in the longer form of presentations. Find

**40**

the time to do multiple drafts of your presentation. If it works for novelists such as John Irving, imagine how it can work for you.

John Irving also offers another important piece of creative advice on another page of his web site: "I always begin with a last sentence; then I work my way backwards, through the plot, to where the story should begin."

Mr. Irving is echoing the advice of what Aristotle called *teleology*— the study of matters with their end or purpose in mind. Today business author Stephen R. Covey stresses the importance of starting with the objective in sight, one of *The Seven Habits of Highly Effective People*, his bestselling book. Whenever you start to develop any presentation, always begin with your end in mind. Decide on your goal or call to action, and then build your presentation with information to support that goal.

Another novelist among those interviewed for the *Wall Street Journal* article had additional advice about the creative process that is applicable to presentations, and you can find her techniques in the next chapter.

# 19

## Presentation Advice from Novelists II
### *Storyboard and Verbalize*

In the previous chapter, you read about an article in the *Wall Street Journal* on the creative processes of novelists that offered two valuable pieces of advice for presenters:

- Begin with your goal or objective in mind.
- Write, rewrite, and rewrite.

That same article provided two more pieces of advice from one of the novelists interviewed, Edwidge Danticat, the author of *Breath, Eyes, Memory,* an Oprah's Book Club selection. Consider Ms. Danticat's first suggestion:

> *Before she begins a novel, Edwidge Danticat creates a collage on a bulletin board in her office, tacking up photos she's taken on trips to her native Haiti and images she clips from magazines ....*
> *[She] says she adapted the technique from storyboarding, which filmmakers use to map out scenes. (F19.1)*

Television and film directors use storyboards to plan their end products, whether it is a 60-second commercial or a multimillion-dollar special-effects epic. They map out the camera angles of each scene and then envision how the individual scenes will combine into a whole sequence. The storyboard provides a 35,000-foot view.

The equivalent in presentations is the Microsoft PowerPoint Slide Sorter, a 35,000-foot view of all the slides in the deck. In the Power Presentations programs, we provide our clients with an electronic (and paper) version of the Slide Sorter view called Storyboard. You can download a soft copy of this form from our web site, www.powerltd.com. Both versions provide a panoramic view of your story.

This view enables you to see all the slides in your presentation at a glance, a perspective that minimizes your focus on details and offers a broader outlook of the landscape. It's an efficient planning tool that helps you check the progression of your story.

You can then validate the progression by speaking your narration aloud with the storyboard in front of you. This practice method is a variation of Verbalization, the subject you read about in Chapter 13, "Do You Know the Way to Spanish Bay?" Ms. Danticat Verbalizes. As the *WSJ* article described it, "She makes a tape recording of herself reading the entire novel aloud ... and revises passages that cause her to stumble."

As part of your preparation, display your slides in a panoramic view and narrate your presentation aloud. Revise your presentation until you are comfortable with the flow. By the time you stand up in front your actual audience, your presentation will be clear, crisp, and fluid.

# 20

## Microsoft Slogans Score a Trifecta
### *Three Persuasive Techniques*

Microsoft's venerable slogans, "Where do you want to go today?" and "Your potential, our passion," are successful because each of them deploys three powerful persuasive selling factors: a call to action for Microsoft, a benefit to its customers, and the most persuasive word in the English language: *you.*

- **Call to Action.** The classic sales technique of asking for the order is usually expressed in taglines or slogans that hard-sell a company, its product, or its service. "Acme: Best of Breed," "Acme: New and Improved!" or "Buy Acme Now While They Last!" These taglines are about the vendor and not about the buyer. Both Microsoft taglines are about the vendor, too. Although they are more soft-sell than Acme's, they still clearly indicate that Microsoft has, respectively, the capability to get its customers wherever they want to go and that Microsoft has the passion to help them realize their potential. However, both of these taglines go one vital step beyond Microsoft itself by involving the buyer—with a benefit.

- **Benefit.** A constant fact of life in business is that most salespersons—thoroughly schooled in their product and enamored with its features—neglect to state its benefits. Ask senior sales managers about their greatest challenge, and most of them are sure to respond that it is to remind their sales force to sell benefits. Some taglines do get it right, as in "Tastes Better," "Costs Less," or "Works Faster." Both Microsoft taglines are infused with benefits. The first indicates that Microsoft's customers can achieve instant gratification, and the second indicates that they can indeed fulfill their potential.

- ***You.*** Chapter 2, "Obama and You," referenced an unsubstantiated Yale University study of persuasive words, with *you* leading the list. Yale never actually conducted such a study. The first unconfirmed and unattributed reference to a similar list is from an ad in the *New York Times* in 1961; only later was the list attributed to Yale—again unconfirmed. Over time, the list has taken on a life of its own, and now it has become a full-fledged urban legend—a vivid example of pre-Web viral marketing.

Unsubstantiated or not, the persuasive power of *you* is undeniable because it addresses the end user of the statement directly. Microsoft involves its audience—the existing and potential customers of its products—with the *you* in "Where do you want to go today?" and in "Your potential, our passion."

The first tagline ran from 1994 to 2002; the second began in 2003 and is still active today. Just before the launch of the latter slogan, the *New York Times* ran a long profile of the company called "Microsofter," in which CEO Steve Ballmer "laid out a new mission statement for the company: 'To enable people and businesses throughout the world to realize their full potential.'" (F20.1)

The statement prompted Steve Bodow, the writer of the *New York Times* article, to comment, "This extraordinarily expansive statement was notable for how little it specifically said about software or computers. Instead, it was about values and corporate culture." Mr. Bodow was describing a soft-sell call to action and a set of benefits.

By mixing those two features with a liberal dose of *you,* Microsoft created two picture-perfect and powerful marketing brands.

Take a lesson from the Microsoft slogans: Ask for the order, offer benefits to your audience, and use *you* as often as you can.

# 21

## Presentation Advice from a Physician
### *Audience Advocacy*

Power Presentations was recently honored by the presence of our first Nobel Prize winner. He is Dr. James E. Muller, the CEO and Chief Medical Officer of InfraReDx, a Cambridge, Massachusetts, company that develops novel photonic-based medical devices to improve the diagnosis and treatment of various diseases. Dr. Muller was one of three American cofounders of the International Physicians for the Prevention of Nuclear War (IPPNW)—the organization awarded the 1985 Nobel Peace Prize.

He and his senior managers participated in a Power Presentations program to develop a financing pitch for InfraReDx. During the program, the team heard—as have countless teams and individuals before them—about the importance of focusing on the audience. We call this Audience Advocacy, a concept that asks presenters to advocate the audience's point of view in equal measure to their own. This focus pervades every aspect of every presentation, starting with the development of the content and continues on to include the design of the graphics, the interaction with audience members, the response to their questions, and even extends to the use of presentation equipment.

At the end of the program, Dr. Muller remarked how Audience Advocacy resonated with the practice of medicine as he had learned it. His mentor had been Dr. Francis W. Peabody, who was instrumental in establishing the Peter Bent Brigham Hospital and the Rockefeller Hospital. During the early part of the twentieth century, Dr. Peabody lectured widely on the subject of physicians and patients. One of his most famous talks concluded with these words:

> *For the secret of the care of the patient is in caring for the patient.*

The prestigious *New England Journal of Medicine* included Dr. Peabody's lecture in one of its publications and commented:

> *These words burned indelibly into the minds of generations of medical students … and the words have lasted well. The lecture, entitled "The Care of the Patient," is reprinted in this book and deserves reading, particularly today, when medical technology focuses more on the disease than on the patient. (F21.1)*

Dr. Muller sent the historic quote to me in an email and concluded, "I would paraphrase it thus for the topic of Audience Advocacy: The secret of care of the audience is caring for the audience."

Concurrent with Dr. Muller's email, another publication provided a further resonance with Audience Advocacy from a sector as far removed from medicine as Venus is from Mars. You'll read about it in the next chapter, "Presentation Advice from a Politician."

# 22

## Presentation Advice from a Politician
### *Audience Advocacy*

Karl Rove, who served as Senior Advisor to President George W. Bush during 2000–2007 and Deputy Chief of Staff during 2004–2007, was widely known as "The Architect" because he was considered to be the power behind the throne of the 43rd president. After he left office, Mr. Rove wrote his autobiography, *Courage and Consequence: My Life as a Conservative in the Fight*, in which he offered political advice analogous to presentations.

The advice was called out in a review of the book in the *New York Times* written by Mark Halperin. As the editor-at-large and senior political analyst for *Time* magazine, and coauthor with John Heilemann of the bestselling *Game Change: Obama and the Clintons, McCain and Palin, and the Race of a Lifetime*, Mr. Halperin understands the essentials of communication. He wrote:

> *Students of practical politics should grab a highlighter and some Post-it notes when reading Chapter 4, entitled "What Is a Rovian Campaign?" Even if one disagrees with Rove's politics (especially if one disagrees with Rove's politics), there are some valuable nuggets about how to run winning campaigns. The most vital rule, Rove says, is often violated: "A campaign must first be centered on big ideas that reflect the candidate's philosophy and views and that are perceived by voters as important and relevant." (F22.1)*

Replace the word *voters* with the word *audience*, and you have sage advice for presentations. This is a further extension of Audience Advocacy, the concept from the prior chapter that asks presenters to advocate the audience's point of view in equal measure to their own.

In Mr. Rove's book, he also took the opportunity—as he often does in his other writing and public appearances—to attack President Barack

Obama. There is no love lost between the president and one of his most constant critics, but in politics, all is fair game. Mr. Obama heeded Mr. Rove's advice about the importance of being "centered" on voters.

During an interview in the *New York Times* about his low job approval ratings, Mr. Obama admitted, "What I have not done as well as I would have liked to is to consistently communicate to the general public why we're making some of the decisions." (F22.2)

Turning to the same "Rovian" nugget that Mr. Halperin highlighted, the president added, "One of the things we've been trying to do is to say, 'Boy, let's get out of here more often.' Just talk to folks and listen to folks so that people get a better sense—not just that we're making smart decisions, but that we're also hearing them and their voices and what they're going through on a day-to-day basis."

Apparently, Barack Obama was also heeding Michael Corleone's advice in *The Godfather, Part II*, when he said, "Keep your friends close, but your enemies closer."

Coda: Concurrent with the release of Mr. Rove's book, I was working with a start-up company (still in stealth mode and, therefore, anonymous) to develop its financing pitch, and I introduced the Audience Advocacy concept to them. At that point, the company's CFO smiled and said, "Sounds like the advice I gave to my 13-year-old son who was about to go on his first date. He asked me what he should talk about. I replied, 'Make it all about her.'"

# 23

## Ronald Reagan Meets Lenny Skutnik
### The Catalyst of Human Interest Stories

On January 13, 1982, during a blinding snowstorm, Air Florida Flight 90 took off from National Airport in Washington, D.C., and suddenly plunged into the Potomac River, killing 74 passengers. Only five people survived the crash. One of them was a woman who owed her life to the courageous efforts of a federal employee named Lenny Skutnik. At the fateful moment, Mr. Skutnik, who was on his way home from work, dove into the icy waters, swam to the woman's rescue. In doing so, he became a catalyst for a speaking technique used by every U.S. president from Ronald Reagan to Barack Obama.

Two weeks after the crash, Mr. Reagan, during his annual State of the Union address, lauded Lenny Skutnik for his heroic effort. In fact, Mr. Reagan had Mr. Skutnik sit next to Nancy Reagan in the congressional gallery. When the president introduced him, Mr. Skutnik rose to thunderous applause from the senators and representatives in the chamber. From that moment on, Mr. Reagan's gesture became a tradition—since followed by George H. W. Bush, Bill Clinton, George W. Bush, and Barack Obama—to recognize ordinary people who had performed extraordinary acts. The tradition became known in Washington as a "Skutnik."

The lesson for presenters is to find examples of human interest stories that help illustrate your own stories. After all, the story every president wants to tell in that annual—and constitutionally mandated—message to Congress and the nation is that the state of the union is strong. What better way to illustrate strength than to personify it with anecdotes of heroic efforts?

What better way to illustrate the strength of your business story than to personify it with anecdotes of satisfied customers, partners, or investors? Find your own "Skutniks."

In a footnote to history, National Airport in Washington, D.C., has since been renamed Ronald Reagan National Airport.

# 24

## Human Interest Stories:
## A Double Advantage
### *Two Ways to Use Anecdotes*

Ronald Reagan was the master of the human interest story that preceded and followed the Lenny Skutnik story that you read in the previous chapter. Mr. Reagan had developed his talent for the human touch in his early days in radio, but he perfected the art of the anecdote during his presidency. Whenever Mr. Reagan spoke, he rarely missed an opportunity to refer to a dedicated student, a brave soldier, or a kind senior citizen, often by name. Presenters would do well to emulate The Great Communicator and intersperse their presentations with brief stories of people who are involved with their business's product or service.

One of my frequent tasks as a presentation coach is to remind my clients to insert such descriptive examples in their presentations. I often do so during a run-through by interrupting their forward progress with the words, "For example ..." This usually brings a smile to the person's face, and they immediately come up with an illustrative case study. More often than not, I have to make this prompt to people in the Life Sciences sector whose companies are involved in either drug development or medical devices. Their tendency is to make a deep dive into their esoteric technology, at which point I interrupt them and say, "Do you have any patients?" This brings that same smile to their faces, and they, too, come up with an illustrative case study.

You can employ such anecdotes in two ways: as supporting evidence for a particular point or to make the human interest story the throughline for the entire presentation.

Let's say that you work for a drug company and you have a patient named John Smith. You can describe the illness John Smith has contracted, how many other John Smiths in the world have the same illness, how much money is spent on all those John Smiths, and how long they've suffered without a cure. Then you can talk about how your company's drug cured John Smith, the patents you have on the drug, its regulatory status, its clinical status, its cost of manufacturing, its average selling price, and its potential profit margin. Finally, you can describe how John Smith was rehabilitated and reimbursed, thus explaining how your drug will sell in the managed care environment. The story of John Smith organizes and humanizes all the details of your company's entire story.

John Smith is every man—and so is your audience.

# Section II

## Graphics: The Correct Way to Design PowerPoint Slides

# 25

## The Presentation-as-Document Syndrome

### *Never the Twain Shall Meet*

In the many years since the publication of the first edition of *Presenting to Win*, I am proud to say that it has made a significant impact upon readers, selling more than 100,000 copies in 12 languages and named by *Fortune* Magazine as one of eight "must read" books. By the same token, I am surprised to say that the book has not had as great an impact upon the presentation trade. Despite the book's extensive reach and the continuing stream of clients who take the Power Presentations program upon which the book is based, most presenters, after reading the book or taking the program, default to a practice counter to the main theory in its pages.

Simply put, that theory is stated in the subtitle: *The Art of Telling Your Story*. True to its promise, the book offers techniques about that classic art of rhetoric, but it does so for only two-thirds of its total pages. The other third of the book is about graphic design in presentations, yet that aspect is not even mentioned on the cover. The imbalance is intentional.

This emphasis on the story—which includes sharp audience focus, clear structural flow, strong narrative linkages, persuasive added value, and even specific positive verbiage—is because the story is more important than the graphics. No audience will react affirmatively to a presentation based on graphics alone. No decisions are made, no products sold, no partnerships forged, no projects approved, and no ships of state launched based on a slide show. Witness the powerful speeches that move hearts and minds: State of the Union addresses, inaugural speeches, nominations, eulogies, sermons, commencements, keynotes, and even locker room pep talks. None of them uses slides.

Therefore, *what* presenters say and *how* they say it are far more important than what they *show*. Does this mean that you should abandon all slides, ye who enter the podium area? Not at all. PowerPoint has become the medium of choice from grade-school classrooms to corporate boardrooms, and far be it from me to advise a sea change as radical as complete rejection of all slides. All I ask—no, urge—is that you use the software properly. Apply the repertory of techniques provided in the other third of the book; the most essential of which is the overarching principle of relegating graphics to a supporting role. Make the presenter—you—the primary focus of the presentation.

Unfortunately, this seemingly simple plea for a shift of emphasis has garnered very few converts. Presentations are still universally defined by and equated with the slides. This forces the slide into two unrelated functions: as a display *during* the presentation and as a record for distribution *before* or *after* the presentation, as handouts. In this fowl/fish (pun intended) double whammy, known as the *Presentation-as-Document Syndrome*, neither version serves its intended purpose, and each version is severely compromised.

If you need a document, create a document and use word-processing software. If you need a presentation, create a presentation and use presentation software. Microsoft Office provides Word for documents and PowerPoint for presentations. Both products are bundled in the same suite, but they are distinctly separate entities. As Rudyard Kipling said about East and West, *never the twain shall meet.*

Use the right tool for the right job. *You* are the storyteller, not your slides.

# 26

## Blame the Penmanship, Not the Pen
### *Operator versus Machine Error*

Edward Tufte is a virtual one-stop shop for all things graphical. He lectures, publishes, writes, and comments extensively on the design of data. One of his books, *The Visual Display of Quantitative Information*, has long been a standard on many business bookshelves. In 2003, Mr. Tufte published a widely distributed 32-page pamphlet called *The Cognitive Style of PowerPoint*, in which he contends that "PowerPoint routinely disrupts and trivializes content" and charges that the software's templates "weaken verbal and spatial reasoning, and almost always corrupt statistical analysis." (F26.1)

This charge is the equivalent of blaming the Montblanc pen company for illiteracy and illegibility.

Yes, most of the "trillion slides each year" that Mr. Tufte derides do indeed abuse PowerPoint's functionalities, but it is the penmanship that is at fault, not the pen.

The equivalent of poor penmanship in presentations is when a user employs PowerPoint as both a document and a visual aid. This multitasking of two distinctly separate functions has become Standard Operating Procedure in business today, as ingrained as a religious ritual. In fact, the document function of PowerPoint has snowballed into multiple subsets, including send-aheads, leave-behinds, and speaker notes. The practice persists because, in the pressured world of business, multitasking and repurposing are equated with efficiency. But these shortcuts disrupt and trivialize the presentation as Mr. Tufte correctly observes. The extra information in presentations gives both presenter and audience too much to process, and in documents, gives the reader too little to process. A PowerPoint deck does not stand alone, and no decision maker will make a decision based on a slide show alone.

The correct role model for graphics exists 24/7 in television news programs. Anderson Cooper of CNN, Katie Couric of CBS, and Brian Williams of NBC provide the details for the stories they tell; the graphics that accompany them are simply illustrative headlines. Professional broadcasting organizations have sophisticated graphical capabilities far beyond those of PowerPoint, yet the images they show are there only as support for the newscaster.

For your presentation graphics, follow the correctly balanced role model you see on all those television news broadcasts. The newscasters tell the story, while the professional graphics that flit by over their shoulders are simply headlines. Emulate Anderson Cooper, Katie Couric, and Brian Williams when you present. Make your slides the headlines, while you provide the details. In the Show and Tell of Presentations, Power-Point is for the show; you do the telling.

Consider bullets as headlines and your discussion as the body text. Consider numeric charts as trends and your narrative as an interpretation of the trends. Consider illustrations as talking points; you add illuminating examples. All the narrative flow and added value must come from you.

For the sake of your audiences, focus on the penmanship, not the pen.

# 27

## You Can't Use a Sentence As a Prompt!
### *Less Verbiage Is More Useful*

A senior engineering manager at a large telecommunications equipment company was one of the participants in a Power Presentations program. True to her technical nature, she wanted to be as accurate in her presentation as in her work, so when she headed up to the front of the room to deliver the pitch she had prepared, she brought along her laptop, with her slides as speaker notes. Having diligently practiced, she started her pitch smoothly, but about a minute into it, she lost track of her content, a fate that befalls many people in front of live audiences. Her eyes darted down to the laptop for a cue, and she suddenly froze. A horrified look came across her face and she blurted, "You can't use a sentence as a prompt!"

In that one moment, she understood the importance of the difference between a presentation and a document. When her eyes moved down to the slide, she saw a full sentence with all the necessary parts of speech: articles, conjunctions, prepositions, and helping verbs. That construct would have been necessary in a document, which must be free-standing and independent of the presenter.

But in a presentation, in which the presenter is the focus and the slides function only as support or illustration, bullets must be treated as headlines, containing only key words: nouns, verbs, and modifiers. Had the engineer followed that simple notion, she would have seen only a few words in her glance, more than enough to serve as a prompt. When she realized that a sentence, with all its detailed parts, is a hindrance rather than a help, she became an instant convert to the headline approach.

Are you ready to convert?

# 28

## Baiting the Salesperson
### *Selling Is about In-Person Communication*

In the pressured world of business, multitasking and repurposing are equated with efficiency. These practices result in the inefficient use of Microsoft PowerPoint for both presentations and documents, with multiple variations of the latter: leave-behinds, speaker notes, uniformity of corporate messaging, and send-aheads.

Of all the many shortcuts, the worst and most common offender is the last: using PowerPoint for both the presentation and a preview of the presentation, as in "Send me your slides in advance." The primary perpetrators of this duality are the solicited, the people in a position to make "yes" or "no" decisions. These power players, who are sought after to become investors in or customers of the presenter's line of business, wield so much weight that their requests are difficult to refuse.

One such decision maker wields his power differently. He is a senior vice president of a Fortune 500 company, and he spends 7 million or 8 million euros every month buying products and services for his organization. Whenever he is solicited by a vendor seeking an appointment, his standard reply is, "I don't have a lot of time for meetings; just email me your slides with information about your product/service and I'll see if I'm interested."

If the vendor agrees, the vice president deletes the email when it arrives, without opening it. His reason: "If a salesperson succumbs that easily, either he can't be a very good salesperson or he hasn't much faith in his own product or service. That person has yielded the selling process to an incomplete impersonal summary. Selling is about person-to-person communication."

The vice president responds differently if the vendor doesn't take the bait. If the vendor offers to send a fact sheet or an executive summary—a true document created in Microsoft Word—as a preliminary to a face-to-face meeting, the vice president accepts. When the meeting occurs, the presentation—created in PowerPoint—functions as an illustration of the presenter's story. Then the vice president can get a complete picture, ask questions, and truly evaluate the salesperson and his or her product or service.

Don't take the bait. Send a document ahead, and bring the presentation along.

# 29

## PowerPoint and Human Perception
### *Scientific Support for Graphics Design*

Dr. Stephen Kosslyn, who chaired the Department of Psychology at Harvard University, has spent 35 years focusing on how the brain recalls visual stimuli in the form of mental imagery. He is also a prolific author, with two relevant works about presentation graphic design: the book *Clear and to the Point: 8 Psychological Principles for Compelling Power-Point Presentations* and an article he wrote with Robert Lane (published on the excellent presentation design site indezine.com), called "Show Me! What Brain Research Says about Visuals in PowerPoint."

In his article, Dr. Kosslyn provided scientific validation of the fundamentals of human perception that are also fundamental in television and cinema—and, therefore, applicable to presentations. Video- and film-makers craft their programs and films on the basic premise that audiences' emotions can be driven by how the human sensory system processes sights and sounds. If directors or editors want to create a pleasant or positive experience, they assemble their audio and video images in smooth, synchronized sequences. If they want to create tension or disturbance, as in suspense, war, chase, caper, or Western stories, they assemble those images in disruptive, asynchronous sequences.

Presenters always want to create only favorable impressions upon their audiences, but in their desire to validate their important ideas, they bulk up their PowerPoint slides with loads of data and jabber away as they click through them. Instead of impressing their audiences, they lose—or, worse, alienate—them.

Dr. Kosslyn explains why: "Viewers must try to read the text, look at the picture, and pay attention to the speaker's words, all in a short time span. Most of us fail to do all three and either: ignore the text and listen to the speaker, or try and read the text and miss the speaker's words." (F29.1)

This disastrous disconnect occurs because the instant a new image appears on the projection screen, the audience suddenly shifts their attention to the screen and away from the presenter, and they do so *involuntarily,* driven by the·reflex actions of their eyes. The audience is so focused on the slide they don't hear anything the presenter says.

Two remedies offer a solution to this slideshow-stopping impasse.

One is Dr. Kosslyn's recommendation, based on his own scientific study, to substitute images for text wherever possible. As he puts it, "Showing people meaningful, content-based visuals, as opposed to text, lessens their cognitive exertion and improves overall experience." (F29.2)

Bravo! Dr. Kosslyn takes *Less Is More* to a new level.

The other remedy lies in the presenter's delivery: Whenever you introduce a new slide or a new element on a slide, whether it's a graphic or text, pause. Stop talking, turn to the screen, and look at the new information. During your pause, look at the image as if you have never seen it, giving your audience time to see it—because they most certainly never have seen it. Your pause fulfills Dr. Kosslyn's goal to lessen your audience's cognitive exertion and improve their overall experience. Only then can your presentation succeed.

Think about that· The key to the effective use of PowerPoint is the pause.

# 30

## PowerPoint Template: Combined Picture and Text
### *The Best Positions for Pictures and Text*

The Microsoft Office Online site offers PowerPoint users a variety of graphical templates for download. One involves combining picture and text in one frame, as in Figure 30.1.

Sample text here

**Figure 30.1    PowerPoint template: text on left, picture on right**

In Figure 30.2, the position of the picture and the text are reversed.

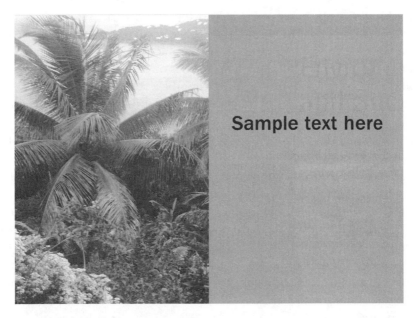

**Figure 30.2   Now picture is on left and text on right**

Now let's up the ante in Figure 30.3 by increasing the amount of text in each picture/text combination and adding four short bullets, as is often done in presentations. (F30.1)

Do you feel the difference?

**Figure 30.3   Adding bullets to the slide**

Most people in Western cultures find it's easier to process the arrangement where the picture is on the left and the text is on the right. Do you? In Western cultures, people are accustomed to reading from left to right. Therefore, when the picture is on the left, your eyes start by taking in the image with one swift scan; then your eyes continue to the right to take in the text by traveling back and forth across each bullet line. In the reverse arrangement, with the text on the left, your eyes start by making all those back-and-forth moves through the text, while aware that there is still another image (the picture on the right) for you to scan.

This juxtaposition, although seemingly slight, makes your audience work harder to take in the images and can accumulate into a negative effect when combined with other thoughtlessly designed or crowded slides. In the previous chapter, you read about the scientific research of Dr. Stephen Kosslyn, of Harvard University, whose advice is worth repeating: "Showing people meaningful, content-based visuals, as opposed to text, lessens their cognitive exertion and improves overall experience."

Follow the doctor's orders. Make it easy on your audience, and they will make it easy for you.

# 31

## Shady Characters
### *The Wrong Way and the Right Way to Build Text*

The default for building text in Microsoft PowerPoint—and the universal practice in presentations—is to dim the outbound bullet by turning it gray, making it almost disappear into the background, as if to say, "I'm done with that item."

Small problem: You're not *quite* done with it, for it is still partially visible to your audience, as in Figure 31.1.

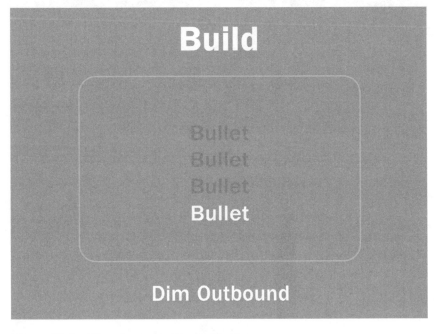

**Figure 31.1   Dimming an outbound bullet**

If your audience wants to refer to one of the earlier bullets, they have to squint to see it. That's a slight discomfort for your audience, but a discomfort nonetheless. If instead you changed the default so that you highlight the inbound bullet with a brighter shade and leave the outbound bullets in their original *visible* contrasting shades, your audience would easily be able to refer back to an earlier bullet and then return to the current bullet. See the difference in Figure 31.2.

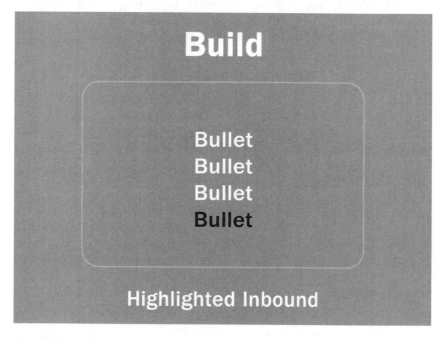

**Figure 31.2   Highlighting an outbound bullet**

Get rid of those shady characters. Presentations are not ghost stories or film noir. Build your case with crystal clarity. Make it easy for your audiences, and they will make it easy for you; the alternative is not a pretty picture.

# 32

## "I Can Read It Myself!"

### *Three Simple Steps to Avoid Reading Slides Verbatim*

During the past 20 years, I have posed the same question to every one of the thousands of participants who have taken the Power Presentations program:

> "How do you feel about presenters who read the words on their slides verbatim?"

I have also posed the same question to the countless businessmen and -women who have sat in the audiences of other people's presentations. Not a single one of them has said that he or she likes the practice. Their responses, usually accompanied by facial expressions ranging from disdain to anger, usually take one of these variations:

> "Why don't you just email it to me?"
>
> "Why bother?"
>
> "I might as well do my email!"
>
> "I'm not a child!"

And the most common response: "I can read it myself!"

The last reaction brings up a note of perspective: The first time anyone ever read to you was to put you to sleep, and thus you—and every man and woman in every audience you will ever face—are forever programmed.

The pervasive practice of reading slides verbatim has drawn derision in many quarters, including that of comedian Don McMillan's YouTube video "How NOT to Use PowerPoint" (see www.youtube.com/watch?v=GB7S-KOJIfE). (F32.1) But reading slides verbatim also raises a follow-up question: "If everyone finds the practice abhorrent, just who does it?" Is there a group of evil phantom presenters giving presentations a bad name?

The fact is, *all* presenters fall victim to this practice at one time or another and are just too embarrassed to admit it. They fall victim because of three even more pervasive practices:

- Use of the presentation as a document
- Lack of preparation and practice
- Text-heavy design

You read several solutions to counter each of these practices in earlier chapters—the most valuable of which is to minimize the verbiage on your slides and to maximize the use of pictorial images—but two additional solutions can help you avoid the dreaded reading verbatim trap:

1. If you do use bullets, treat them as headlines not sentences, and limit them to four or five words per line. Construct each line to begin with the same part of speech, preferably action verbs. This approach will make it virtually impossible for you to read verbatim.

2. When you present each bullet, paraphrase, use synonyms, or juxtapose the order of the words as you speak them. The audience will be able to extrapolate the meaning of the bullet without having to think, "I can read it myself!"

# 33

## A Case for Case I: Initial Caps or All Caps

### *Text Design in Presentations*

An article in the *New York Times* reported on a trend among major corporations to update their brand logos and noted that several of the companies—among them Wal-Mart, Kraft, Stop & Shop, and Cheer—have done so with "striking similarities." One of the similarities is the conversion from all-caps to initial caps. The *Times* article described this shift as follows:

> *Toned-down type. Bold, block capital letters are out. Their replacements are mostly or entirely lower case, softening the stern voice of corporate authority to something more like an informal chat.*

The article then went on to propose two reasons for the shift. First, the influence of email and text messaging which, like e.e. cummings' poetry, is often composed in all lower case. Second, the long economic downturn has prompted a new look that is "non-threatening, reassuring, playful, even child-like. Not emblems of distant behemoths, but faces of friends." (F33.1)

Should you make a similar shift in the text in your presentation graphics design? Perhaps, but first consider the different circumstances of each medium. A corporate logo is an inanimate object, regardless of the font, colors, or decorative ornaments used. The image sits inert on the product package or in a print, electronic, or video advertisement, from where it must convey a complete, immediate, and consistent message. Conversely, a presentation is a dynamic event in which all graphics, including text, serve only to support the presenter. In this model, the graphics play a secondary role—as a headline, leaving the presenter to provide the details in his or her narrative.

As you read in Chapter 26, "Blame the Penmanship, Not the Pen," the ideal model for the relationship between slides and the presenter is that of television news broadcasts. In that medium, the graphics that accompany any story consist of a simple impressionistic image and very few words of text. The words in the text are usually all-caps as a headline, leaving the newscaster to provide the details of the story. In newspapers and magazines, headlines are often in all-caps, and the body text is in initial caps. Initial caps, as in the illuminated first letter of ancient manuscripts, signal the beginning of a long read.

Headlines are intended to capture attention, and all-caps demands attention. The two most ubiquitous and important attention-demanding signs in the world are "STOP" and "EXIT." However, because all-caps are more difficult to read than initial caps, if you decide to use them in the text in your presentations, keep that text to four or five words, at a maximum.

Therefore, whether you use all-caps or initial caps—the ultimate choice is individual taste—be sure to default to the headline approach. Do as the television news anchors do: Provide the body text to the headline on the presentation screen.

# 34

## A Case for Case II: Serif or Sans
### *Font Design in Presentations*

All the typeface fonts available to your presentation graphics fall into two major categories:

- Serif, in which the letters have decorative hooks at the ends of the strokes. The most common serif fonts are Times New Roman, Garamond, and **Cambria**.
- Sans serif, in which the letters have only straight or round strokes, with no decorative flourishes. The most common sans serif fonts are Arial, Calibri, and Century Gothic.

The little hooks in serif font make text easier to read because they enable a reader's eyes to distinguish individual letters. Please note the feeling in your eyes in the serif and sans serif versions of the same sentence:

The quick brown fox jumped over the lazy hen.

The quick brown fox jumped over the lazy hen.

See the difference? Does that mean that serif is preferable? Not necessarily. First and foremost—as always—consider the *Less Is More* principle for presentations. The differential in fonts is diminished when you reduce a sentence to a headline, as *all* text in *all* presentations must be treated. A headline uses only key words, primarily verbs and nouns, with very few articles, conjunctions, prepositions, and adjectives, the parts of speech that form a complete sentence. Please note the headline version of the previous sentence in both serif and sans serif:

Fox jumped over hen

Fox jumped over hen

The four-word headline conveys the same basic information as the earlier nine-word sentence, and the brevity reduces the difference between the serif and sans-serif versions. Each version of the headline is easy to see and grasp.

In a presentation, the presenter would discuss the fox's quickness and the hen's laziness. In a document, which must stand alone, the descriptive adjectives and the articles are necessary for narrative clarity. For visual clarity, however, serif fonts make the full sentences easier to read.

Another factor to consider in choosing between serif and sans serif is popular usage. In the previous chapter you read about two of the most ubiquitous and important attention-demanding signs in the world: EXIT and STOP, each of which is composed in sans serif.

Finally, consider personal taste and the Latin phrase *De gustibus non est disputandum*—that is, "There is no argument about taste."

# 35

## What Color Is Your PowerPoint?

### *Contrast Counts*

The previous chapter about serif and sans serif font concluded with the Latin phrase *De gustibus non est disputandum,* or, "There is no argument about taste." The phrase is even more applicable, if not indisputable, when it comes to color choice. Well, *almost* indisputable, for there is a single unavoidable consideration that transcends the taste of any presenter or presenter's designer: the audience's ability to understand the graphic.

A simple one-word rule, applicable to every element of every graphic, will make it easy for every audience to understand your every slide. That one word is *contrast*. A simple way to implement contrast is to reference the classic color wheel, which is divided into two halves: warm colors and cool colors. The warm colors are the yellow/orange side of the wheel, and the cool colors are the blue/green side. By choosing a warm color as the background and a cool color as the foreground, or vice versa, you achieve contrast by default.

Yellow against blue or blue against yellow provide one of the sharpest contrasts you can create, a fact borne out by the U.S. Navy. During World War II, when aircraft carriers came into large-scale use, the greatest challenge was landing the fast-moving warplanes on the decks of the bobbing ships; a challenge heightened by the limited visibility of the blue of the sky and the blue of the sea that framed the gray ships. To assist the pilots as they approached the carriers, a flight director stood on the deck's landing strip and gave visual signals. These flight directors wore yellow life vests that made them stand out clearly against the blue sky and water.

(The colors of the life vests of other personnel were chosen more for relevance than contrast: red for ordnance, white for medical, green for maintenance, purple for fuel, and blue, the least contrast against the sea and sky, for personnel who handled the planes *after* they landed.[F35.1])

Create your slides with sharp contrast between the foreground and background. In all the Power Presentations programs, our slides follow the fleet, with bright yellow foreground text that stands out clearly against a royal blue background.

Use any cool color/warm color foreground/background combination for your presentation. Or use white text against any dark background or black text against any light background. Boldface and shadowing your text sharpens the contrast even more. In fact, if you enclose your text in a shape, use a drop shadow on the shape to give it contrast.

One other important factor to consider regarding contrast is the gradient feature. Today's PowerPoint and other graphics applications provide presenters with many bells and many whistles to create attractive slides. One of the most readily available and frequently used features is gradient shading. Many presenters and their designers, in their desire to prettify their slides, incorporate a gradient in their design. Admittedly, this feature provides attractive graphical texture, but it makes some elements hard to discern. As you can see in Figure 35.1, the first bullet does not contrast with the darker part of the gradient and is difficult to read. If your audience wants to look back that bullet, they would have to squint.

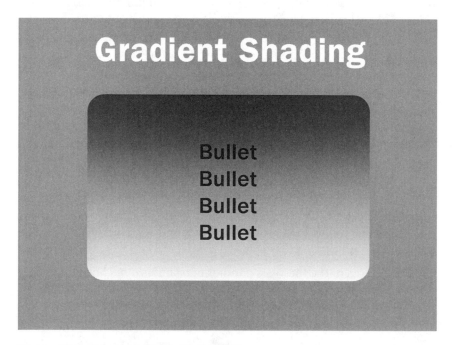

**Figure 35.1    Beware of Gradient shading.**

So let's conclude with the converse of a phrase you've seen often in this book:

Make it hard for your audience, and they will make it hard for you.

# 36

## Presentation Advice from Corona Beer
### *Peripheral Vision Counts*

A delightful Corona beer video commercial, set in their now-familiar tropical seascape, makes a humorous but telling point about peripheral vision. A man and a dark-haired woman are seated in beach chairs with their backs to the camera, facing straight ahead toward the surf. The man is on the left, the woman on the right; between them is a low table on which two bottles of Corona beer stand, each topped with a wedge of lime.

After a moment, a tall and tanned, willowy blond girl wearing a tiny white bikini enters the scene from the right and slowly crosses to the left. The man's head turns and follows the blonde until she leaves the frame. After she has gone, the man's head returns to face front. After another moment, the brunette's hand reaches up, takes the lime wedge from the man's bottle, and squirts it at the man's face. During the entire scene, her head never turns.

The commercial was so successful that Corona produced a follow-up. The setting and the positions of the man and woman are the same as in the first version. In this version, the beer bottles are sitting in an ice bucket and are capped. After a still moment, an attractive muscular young man enters the scene from the left and slowly crosses to the right. This time, the woman's head turns and follows the young man leaving the frame. As she looks off, the seated man reaches into the ice bucket, picks up the bottle closer to the woman, shakes it vigorously, and then replaces it in the bucket. The woman's head then returns to face front, but her arm reaches out to the ice bucket and takes the bottle closer to the man, leaving the shaken bottle for him. Then her arm reaches out again and extends a bottle opener to the man.

In addition to the classic jealous battle-of-the-sexes triangle, the commercial plays on the theory that women have peripheral vision, whereas men have tunnel vision because of our origins as cave dwellers. The widely held theory (you can find thousands of references on the Web) is also expressed in the book *Why Men Don't Listen and Women Can't Read Maps: How We're Different and What to Do About It*, which posits that primitive men, as hunters, had to be narrowly focused on their prey, whereas primitive women, as nurturers, had to have a wider scope of vision for the safety of their children.

Be that as it may, all men and all women share one characteristic regarding their vision: the hair-trigger reflex of the eyes to visual stimuli. Whether in tunnel or peripheral vision, all human eyes react involuntarily to new images. This immutable fact plays a critical role in presentations.

The instant a new graphic appears on a presentation screen, the eyes of every person in every audience immediately dart over to look at it—involuntarily. At this very same moment, most presenters continue speaking. Because the audience's eyes are more sensitive than their ears, they focus on the graphic and lose the presenter's words. The audience stops listening.

If instead, the presenter pauses and gives the audience time to absorb what they see, the presenter maintains the audience's attention.

Combine Corona beer with the classic Coca-Cola slogan, and give your audience the pause that refreshes.

# 37

## The Cable Crawlers
### *How Television Animates Text*

In their drive to feed the insatiable 24/7 monster, cable news channels fill their screens surrounding the central news story with an array of other features displayed in dazzling but sometimes distracting graphics. The visual inclusions (which often become incursions) consist of all or some of these elements: time, weather, sports scores, traffic, stock reports, captions, program promotion, and logos.

One constant element common to all these channels is the crawl, the running banner of news blurbs that streams across the bottom of the screen like the old stock market paper ticker tape. The feature had its electronic origins in the running headlines that wrapped around the historic *New York Times* building on Times Square. The newspaper has long since moved its offices to a new building, but the crawling headlines have spawned countless clones that live on in many other Times Square buildings that make up the tourist spectacle that is Broadway.

On the cable channels, the crawlers follow the way we read text in Western cultures: They enter from the right, travel across the screen, and disappear on the left. However, CNN changed its format. No longer does their text crawl; instead, it appears as a short single blurb of white text that rolls up into a black slot at the bottom of the screen and then rolls out at the top of the slot, replaced by the next text blurb. Because viewers can better take in an entire blurb in one glance than they can while having to follow the text in constant motion, the difference is easier on the eyes.

See for yourself. Look at the ticker-tape crawls on CNBC, MSNBC, and Fox News, and then look at the CNN style. The latter is much easier on the eyes.

When you use animation in your presentation, avoid the traveling options, such as Fly, Peek, Ascend, and Descend, each of which blurs the words as they enter. Instead, use the Wipe option in both Custom Animation and Slide Transition; it brings on the words the way we are accustomed to reading them in print.

Then think about how your audiences react to your graphics. Make it easy for them, and they will make it easy for you.

# 38

## Computer Animation
### *Three Simple Rules*

We've all been in the audiences of far too many presentations that unleash all the bells and the whistles of PowerPoint animation with a frenetic, pyrotechnic display that challenges a Fourth of July celebration or a night at Disneyland.

That such excess happens is no surprise. The many options in the pull-down menus and ribbons of PowerPoint animation are as fascinating as are all the many joystick and button options on the keyboard or controller of a computer action game. Slide Transition alone has 58 effects grouped into 5 categories, with 3 speed options for each. They cry out, "Try me!"

Uncontrolled, they can cause the loss of the game or the presentation.

The obvious solution is to exercise restraint, but that is negative advice. What to do instead? Three simple, overarching rules will bring your presentation to life (after all, that *is* the definition of animation) and, more important, bring clarity, if not tranquility, to your audiences.

**Rule One: Make the default direction of your animation left to right.** Text in Western languages is printed from left to right. This simple fact drives how humans perceive visual stimuli. When your audience sees images move from left to right, it will feel natural and pleasing to their eyes—and make them more receptive to you and your message.

**Rule Two: Use motion to express the action in your message.** If you want to show rising revenues, have your animation move from the bottom up; if you want to show declining costs, have your animation move from the top down. If you want to send a negative message (say, about your competition), reverse direction and move your images right to left.

**Rule Three: Allow your audience to absorb your animation.**
The sensitive optic nerves in your audience's eyes cause them to react involuntarily to light and motion. Therefore, the instant your animation starts, all their attention suddenly shifts to the screen and away from you. Because they are so focused on the animation, they don't hear anything you're saying, nor do they see what you're doing. Therefore, whenever you introduce animation, stop speaking, turn to the screen, and allow the animation to complete its full course of action.

Think of these three rules as using animation to tell your story just as a Walt Disney movie does, but leave the fireworks to Disneyland.

# 39

## PowerPoint and the Military
### *Sometimes More Is More*

A *New York Times* article by Elisabeth Bumiller, titled, "We Have Met the Enemy and He Is PowerPoint," showed a dense, complicated PowerPoint slide that the Pentagon used in a presentation to describe the complexity of the situation in Afghanistan. The slide, a swirl of overlapping lines, arrows, words, and colors, resembling a bowl of spaghetti, proceeded to make its viral way around the Internet as yet another example of the abuses of PowerPoint, particularly by the military. (F39.1)

Richard Engel, the NBC News chief foreign correspondent, was the first to publish the slide on his blog, where he described two diametrically different reactions to the slide from the military itself:

> *For some military commanders, the slide is genius, an attempt to show how all things in war—from media bias to ethnic/tribal rivalries—are interconnected and must be taken into consideration .... But for others, the diagram represents a fool's errand that the United States has taken on in the name of national security. (F39.2)*

Another reaction came from General Stanley A. McChrystal, the then—and since summarily deposed—leader of American and NATO forces in Afghanistan. According to Ms. Bumiller, General McChrystal said, "When we understand that slide, we'll have won the war."

Ms. Bumiller then went beyond the slide itself to discuss the chronic use and abuse of PowerPoint in the military, quoting two other senior officers: Marine Corps General James N. Mattis, the Joint Forces commander, who said, "PowerPoint makes us stupid," and Brigadier General H.R. McMaster, a veteran of combat in Iraq, who said, "It's dangerous because it can create the illusion of understanding and the illusion of control."

PowerPoint is a constant source of chatter in the blogosphere and the presentation trade, all of it a loud hue and cry against the software in general and its effect on the military in particular. In the vanguard of the attackers is Thomas X. Hammes, a retired Marine colonel, who wrote a widely distributed article, called "Dumb-Dumb Bullets," first published in *Armed Forces Journal.* He complained:

> *PowerPoint is not a neutral tool—it is actively hostile to thought-ful decision-making .... Instead of forcing officers to learn the art of summarizing complex issues into coherent arguments, staff work now places a premium on slide building. (F39.3)*

As you've read throughout this section, I am in violent agreement with Mr. Hammes and all the other critics about the abuses of Power-Point. My plea for simplicity in design is echoed by a cottage industry of consultants, coaches, designers, and authors. Chief among these are Garr Reynolds, whose *Presentation Zen* imaginatively applies the princi-ples of Japanese minimalism to graphic design, and Nancy Duarte, whose *slide:ology* offers a wealth of simple but creative ways to display ideas visually.

However, I depart from counseling simplicity in the case of the Pen-tagon's complex Afghanistan slide and side with those military command-ers who see it as genius. The spaghetti-like image effectively illustrates the complexity of that situation. PowerPoint's most basic function is to illustrate, not to make coherent arguments—that is the role of the pre-senter. Nor is it to document—that is the role of a text file. The one and *only* role of PowerPoint is to show what the presenter is telling.

One of the most illustrative and effective slides I have ever seen in the business world was for the IPO road show of a computer company. It was a single slide that showed the logos of the company's customers—all 600 of them—an image to loosen the purse strings of any skeptical investor.

Sometimes, More *Is* More.

# Section III

## Delivery Skills:
## Actions Speak Louder Than Words

# 40

## The Art of Conversation
### *Eye Contact and Interaction Start at Infancy*

A book called *The Art of Conversation* states its case very clearly in its subtitle: *A Guided Tour of a Neglected Pleasure.* As the *Publishers Weekly* blurb notes, author Catherine Blyth's main focus ranges from "small talk to pillow talk, from riotous raconteurs to crashing bores, from flattery to false smiles."

But Ms. Blyth goes beyond the frivolous to touch on some of the more substantive aspects of human communication. One of them is at the origin of interpersonal exchange. She looks back at how parents relate to infants and says:

> *"Goo-goo" is the most important word in the world, because when parents coo at babies, they're educating them in what behaviorists call "musical companionship." As babies goo-goo back, they absorb timing, taking turns, tone, coordination, gestures, facial expressions, storytelling—the orchestra of instruments by which emotions are transmitted and relationships formed. (F40.1)*

Ms. Blyth's description of early relationships resonates with that of another book called *Bonding: Building the Foundations of Secure Attachment and Independence*, by neonatologists Marshall H. Klaus, John H. Kennell, and Phyllis H. Klaus. Although *Bonding* is more than two decades old, it has become a classic for new parents. In it, the doctors report on their studies of the behavior of mothers and their newborn infants. One of the primary dynamics they explored is eye-to-eye contact. Marshall Klaus noted:

> *The visual system provides one of the most powerful networks for the mediation of maternal attachment. The babies who received this kind of attention went into a state called "quiet*

*alert," in which, rather than squirming and looking around as babies are apt to do, they remained focused on their mothers. (F40.2)*

Taken together, both books describe the early imprinting of visual and vocal expressiveness and interaction in human communication. These dynamics carry into the adult life of every man and woman. Presenters can leverage these powerful forces with their own "orchestra of instruments."

Treat every presentation as a series of person-to-person conversations. Make eye contact, be expressive and engage with every member of every audience.

In the next chapters you'll learn the many benefits of this fundamental but powerful technique.

# 41

## Presentation Advice from Edward R. Murrow
### The "Person-to-Person" Role Model

Edward R. Murrow, the celebrated broadcast journalist, had a distinguished career at CBS that started in the early days of radio in 1935 and culminated in the golden age of television in 1961. George Clooney's 2005 film *Good Night, and Good Luck* brought the man and his era back to the forefront of public consciousness, but I didn't need any reminders. My own career at CBS began after Mr. Murrow had left, but his legend was still reverberating throughout the studios and control rooms of the Broadcast Center on 57th Street in New York when I got there.

One of Mr. Murrow's major assignments was as the host of a television series called *Person to Person*. Every Friday night for eight years, he sat in a studio in Manhattan and engaged in casual conversation with a different celebrity. But their conversations were only virtual; the celebrities were in their own homes at remote locations. Their images were transmitted over coaxial cables and composited with that of Mr. Murrow using a then-novel technology called chroma-key. His talent was to bridge all the technology and make all the conversations natural.

As a staff producer of unsponsored public affairs programs at the local station, I didn't have quite Mr. Murrow's network clout or his advertisers' dollars, but we did use the same staple television format: the talk show. With our meager budget, all the participants were in the same studio, but the format provided an important benefit.

None of the guests on our public affairs programs—who came from the business, government, academic, and scientific sectors—was a professional performer. My job was to help make those men and women feel comfortable and look comfortable in the highly stressful circumstances of appearing on camera. To do this, we leveraged the talk show format.

We structured our programs as conversations, as either person-to-person interviews or small group discussions conducted by professional moderators. By putting our nonprofessional guests into familiar settings, we reduced their stress levels. (Of course, once we got our guests to feel comfortable, we made them feel *uncomfortable* by having our moderators ask tough questions. But that is another subject for the next Section, "Q&A: Handling Tough Questions.")

Years later, when I left television and became a speaking coach, I wound up with essentially the same client base as I'd had in public affairs television. So I used the same approach in business—treating presentations as person-to-person conversations—and produced the same calming results as I did in television. In doing so, I was able to help alleviate one of the most common apprehensions known to humankind: the fear of public speaking.

Although I never had the privilege of meeting Edward R. Murrow, his legacy extended, over the coaxial cables of time, to inform and inspire the Power Presentations methodology.

# 42

## Nonverbal Communication
### *Look Them in the Eye*

Major newspapers, along with countless other print publications, have focused on what is rapidly becoming their chief competition (and perhaps successor): electronic communication, especially electronic social networking. Text messages, emails, blogs, Facebook, and Twitter, although lacking the organizational breadth and professional depth of newspapers, have clearly encroached on print territory.

One of the major side effects of this cultural change is in the area of interpersonal communication, particularly among young people. A *Wall Street Journal* article, "Why Gen-Y Johnny Can't Read Nonverbal Cues," focused on these dynamics. The article described "a culture where young people—outfitted with iPhone and laptop and devoting hours every evening from age 10 onward to messaging of one kind and another—have resulted in a 'text-dependent world' in which nearly all of their communication tools involve the exchange of written words alone." (F42.1)

This might be all well and good among socializing Gen-Yers, but it won't fly in business. The single most important factor in making business decisions is how people communicate face-to-face. Every business exchange has two sides: the solicitor, a person in search of an objective—a sale, an investment, a partnership, or a project approval—and the solicited, a person who can approve or reject the solicitor's objective. From the most basic business exchange (a job interview) to the most challenging (an investment), the ultimate decision is always based on how well the solicitor communicates his or her message to the solicited—in person.

The most telling example of this dynamic is in the investment arena, and it occurs at the interface between interpersonal and Internet communications. When a company decides to sell shares to the public for the first time, it files a formal Initial Public Offering (IPO) with the U.S. Securities and Exchange Commission. An important adjunct to the offering is a road show: a presentation that the company's senior management delivers in more than a dozen cities, presenting 6–8 times a day for a total of 70–90 iterations, all within two intense weeks.

Since 2005, all IPO road shows are now recorded and posted on a web site called retailroadshow.com. Anyone with a browser can view any company's archived presentation displayed in a split screen. One side shows a video of the senior managers presenting, and the other side shows their slide show, clicking along with the narrative. Despite this universal accessibility, offering companies still send their senior management out on the road for those two intense weeks to deliver those 70–90 iterations, just as they did before retailroadshow.com came into being.

The reason for this grueling tour is that no investor will make a decision to buy millions of dollars of stock (as are most such IPO transactions) based on a canned presentation alone. Investors want to meet the senior executives in person, press the flesh, look them in the eye, and interact with them directly.

I was privileged to coach the IPO road show of a Massachusetts technology company called A123Systems. Before the tour, the investment bankers expected to price the offering somewhere between $8 and $9.50 a share; after the road show, they priced the shares at $13.50, for a total deal value of $380 million. At the end of the first day of trading, shares were at $20.29.

Dave Vieau, the A123 Systems Chief Executive Officer, and Mike Rubino, the company's Chief Finance Officer, did their full two weeks on the road, meeting the investors in person, pressing the flesh, looking them in the eye, and interacting with them directly.

Person-to-person counts big-time in the Big Time ... and every time *you* present.

# 43

## Presentation Advice from Pianist Murray Perahia
### Concentration Creates Control

Murray Perahia is one of the most acclaimed classical pianists in the field. Along the career path to his acclaim, Mr. Perahia was fortunate to have had help from several giants of classical music; the most influential was his teacher, the legendary pianist Vladimir Horowitz. But according to an article in the *Wall Street Journal*, "One thing Mr. Perahia seems not to have absorbed from Horowitz is the latter's legendary stage fright." (F43.1).

Stage fright is, indeed, the stuff of fearsome legend, affecting the public appearances of musicians, actors, and our particular area of interest—public speakers. The cold, clammy hand of dread that grips speakers with such paralyzing power is so pervasive that it has given rise to an entire industry of providers offering relief. If you search the Internet for the fear of public speaking, you'll find millions of entries offering remedies.

Although there are a great number and variety of solutions, the problem remains unsolved because the vast majority of them are purely physical solutions to what is *not* a purely physical problem. The fear of public speaking is caused by a presenter's fear of failure. So unless that mental fear is allayed, physical cures will not work.

Mr. Perahia agrees that "it has to do with a fear that you might fail in some significant way. But it's not something I've spent a lot of time thinking about, because you just have to do it. Communication is a very strong part of art. And to get it, one needs to play it, to live it." As the article on Mr. Perahia concludes, "[A]udiences typically savor his legendary concentration and unassailable technique in hushed form—granting him a degree of respect not always afforded other, equally famous artists."

To bring concentration from the concert stage to the podium, use The Mental Method of Presenting. Succinctly stated, the method involves shifting your mental focus. Don't think about how you are doing—whether you succeed or fail—but on how your audience is reacting to you. You can then respond to what you observe by either pausing to adjust your content or moving forward. This simple shift of concentration gives you control of your own destiny and, in doing so, reduces your fear of public speaking.

I recently met a young businesswoman who, upon learning that I am a presentations coach, proceeded to confide in me and describe her perpetual problem with nerves. She said that whenever she has to face an audience, she goes to the front of the room clutching a stack of note cards and shuts her eyes for before speaking. I suggested that instead she open her eyes and read her audience instead of her notes. That mere summary of the Mental Method brought an immediate sigh of relief and a smile to her face.

Imagine what will happen when she—and you—put the correct focus into practice.

# 44

## Presentation Advice from Actress Tovah Feldshuh

### Concentration Creates Communication

Rudyard Kipling wrote his classic poem "If ..." to commemorate a war hero in the Boer War of 1899–1902. The poem, which begins with "If you can keep your head when all about you are losing theirs ..." and ends eight stanzas later with "Yours is the Earth and everything that's in it," (F44.1) is a paean to concentration.

A *Wall Street Journal* story about rude behavior from audiences at Broadway shows provides a modern example in the theater. The story describes the all-too-common discourtesies of mobile phone ringing and loud talking. But the worst incident was at a performance of *Irena's Vow*, a serious drama about the Holocaust, where a "man walked in late and called up to actress Tovah Feldshuh to halt her monologue until he got settled." The article reported that the actress complied but that "she doesn't recall the incident, which she says may be evidence of the Zen attitude she's cultivated onstage." (F44.2)

Ms. Feldshuh was in a state many actors achieve or aspire to achieve when they are onstage. Call it "Zen," call it "The Zone," "Being in the Moment," call it what you will; that state of total concentration is what makes for great acting.

This is also the state that presenters and speakers would do well to enter, with one important difference. Performers such as Ms. Feldshuh direct their concentration inward, to their characters. To be effective, presenters should direct their concentration outward, to their audiences, and, more specifically, to individual members of their audiences.

The rationale for this shift strikes at the heart of the most powerful challenge to effective presentations: the fear of failure and its accompanying performance anxiety, which are all manifested in one insistent, stressful thought: "How am I doing?" This singular focus causes presenters to turn inward, which only heightens their anxiety.

Instead, if presenters focused on their audiences to see how *they* are doing, the presenters would be able to gauge the effect of their words and react to what they have observed. If they saw their audiences nodding their heads, indicating understanding, presenters could move along; if their audiences appeared to be puzzled or unconvinced, presenters could adjust their content until they got those desirable head nods.

This proactive approach produces a double benefit: It reduces the presenters' performance anxiety, and it provides instant gratification to their audiences—an essential element for success in any presentation. Therefore, concentration not only creates control, but closes the loop to create communication as well.

# 45

## Presentation Advice from Michael Phelps and Dara Torres
### How to Control Stress under Pressure

Michael Phelps won eight gold medals at the 2008 Olympics in Beijing, and Dara Torres won three silver medals. Yet both swimmers had to deal with the inevitable butterflies (pun *somewhat* intended) that go along with such high-pressure events.

These champions have training resources and regimens far above and beyond the scope of any business presenter, but their stress levels are no different. All human beings experience the same adrenaline rush. Mr. Phelps and Ms. Torres have developed some simple and effective anxiety-controlling techniques that they described to the *New York Times*—and that every presenter can use.

Ms. Torres, who captured attention of the world by becoming the oldest athlete to qualify for the Olympics at the ripe old age of 41, described her stressful experience in her first Olympics in 1984 when she was just 17: "I freaked out when I walked out on the pool deck and saw 17,000 people."

One of the veteran members of her team brought Ms. Torres back to the Olympic village and "plopped her in front of a television to watch soap operas to take her mind off her race." This is known as **redirection**, and although presenters cannot step away from the podium and watch soap operas, they can redirect their attention from their own stressful feelings to observe their audience's reactions. In response to what the presenters see, they can adjust their content. If they see that the audience is nodding their heads—reacting positively—presenters can move forward; if not, they can adjust their narrative. This simple mind shift reduces tension.

Mr. Phelps uses two stress-reduction techniques: **structured relaxation** and **visualization**. The first is a progressive relaxation of the individual parts of the body to eliminate the tension. The second technique focuses the mind on the endgame—winning. The *Times* article describes how Mr. Phelps brings the techniques into play. "Once he has cleared his mind and loosened his limbs, Phelps will swim each race over and over in his mind. It is not just the perfect race that Phelps pictures. He sees himself overcoming every conceivable obstacle to achieve his goal time." (F45.1)

Because presenters must remain in the moment and think about their audience instead of their own body parts, structured relaxation has no direct equivalent. But every presenter would do well to pause for a moment before stepping up to the front of the room and take a few deep breaths. Breathing brings oxygen into the body, and oxygen brings relaxation.

The presentation equivalent of Mr. Phelp's form of visualization occurs during the all-important preparation process. Think about your audience in advance; think about what they know and what they need to know to respond favorably to your pitch; think about their goals, desires, issues, concerns, and hot buttons; and think about the questions they might ask. As Michael Phelps puts it, "I do go through everything from a best-case scenario to the worst-case scenario just so I'm ready for anything that comes my way." If you do this in your preparation for your presentation, you will find few surprises when the moment of truth arrives.

Redirection, relaxation, and visualization adapted to presentations will help control the butterflies. Butterflies in the stomach are as common at the podium as they are in the swimming pool. Yet the solution in both venues is the same: Make your butterflies fly in formation.

# 46

## Presentation Advice from Frank Sinatra
### *The Art of Phrasing*

In one of Frank Sinatra's compilation albums, the liner notes contain a meaningful quote from the celebrated singer: "The reading of a song is vital," he said. "The written word is first; always will be. Not belittling the music, but it really is a backdrop. To convey the meaning of a song, you need to look at the lyric and understand."

Imagine that: Frank Sinatra, whose rich, resonant voice was lauded and listened to perhaps more than any other in the history of music, considered the lyrics of a song more important than the melody. The man who was aptly nicknamed "The Voice" relegated his legendary "chops," as professional musicians call it, to second place.

The presentation advice here is an analog of the relationship between a presenter's narrative and the slides. A presenter's words must be positioned, as in Mr. Sinatra's hierarchy, "first; always will be"; the slides must be relegated to what he refers to as a "backdrop." PowerPoint must take second place to the narrative. This hierarchical positioning extends to how the words of the narrative are spoken—their cadence.

Mr. Sinatra's legendary creative phrasing reinforced his primacy of words. Take one of his big hits, "The Way You Look Tonight," written by Jerome Kern, with lyrics by Dorothy Fields. The song, which has been recorded by many other singers, was originally written for the 1936 film *Swing Time* and even won the Academy Award for Best Original Song that year. In the film, Fred Astaire, accompanied by an unseen orchestra of gentle, lilting strings and woodwinds, sat at a piano and crooned the lyrics to the object of his love, Ginger Rogers.

As any love song should, "The Way You Look Tonight" is crafted with soft consonants and many vowels, which serve as a vehicle for long-held notes and lilting phrases. Mr. Astaire interpreted it just that way, softening the sound even further by linking the words. He made each line a long, smooth phrase.

Mr. Sinatra's version was different. He chose an upbeat interpretation and backed it with Nelson Riddle's jazzy arrangement, replete with brass and percussion. To emphasize the beat, Mr. Sinatra did less linking, punctuating the words by articulating the consonants sharply, and inserting abrupt pauses. He made hard stops after the words *keep* and *look*, popping the *p* and *k* sounds, respectively. But Mr. Sinatra also let his rich baritone shine through by holding his vowels even longer than did Mr. Astaire. The *o* in the word *lovely*, the *ea* in *please*, and, of course, the *o* in *love* all rang out with sustained resonance. To paraphrase another line in the same song, with each word, Frank Sinatra's talent shows.

A presenter is not a singer or a performer, and this is not to suggest that you score and arrange your speech to the precise degree that Mr. Sinatra did with his lyrics. But you should develop a crisp, clear, and natural cadence that makes it easy for your audience to follow. We've all been in audiences to presenters who ramble or race, or whose speech pattern is choppy, punctuated by distracting "ums" and "ahs." It's like listening to a singer who sings off-key and/or off-beat.

To develop your cadence, listen to your own speech pattern objectively. Use the record function on your smartphone or a dictating recorder to capture your voice while you're speaking on a landline or during a meeting. Later, play back the recording and listen to your cadence. Are you rambling or choppy? Do you race? Do you use "ums" or "ahs"? How do you form your phrases? Do you *ever* pause?

The key is to control your cadence, and the process begins by hearing yourself as your audiences hear you. Then you can develop your own natural style—and make the words of your presentation sing.

# 47

## Presentation Advice from
## Soprano Kiri Te Kanawa
### *The Importance of Breathing*

In 1982, soprano Kiri Te Kanawa was awarded the title Dame Commander of the British Empire for her lifetime achievements as one of the leading operatic singers in the world. These days, Dame Kiri spends most of her time in a backstage role with her own foundation, dedicated to helping New Zealand singers and musicians develop their careers.

As an extension of her supportive role, she came to the United States to deliver a Master Class at the San Francisco Conservatory of Music. Seven fortunate students had the opportunity to be coached directly by Dame Kiri. One by one, she listened to each of them sing an aria. She then made suggestions about their musical interpretations. Finally, she demonstrated the same passage with her own magnificent voice, providing an enormous treat for both the students and the audience.

The master gave her disciples invaluable techniques about phrasing, enunciation, posture, and breathing—especially breathing, which is so vital to any opera singer. On several occasions, Dame Kiri went right up to some of the students, placed her hand on their stomachs, and asked them to sing a passage again. When they got to an important note, the teacher pushed hard, forcing their breath and enhancing the note significantly.

Although this teaching device is useful for singers, presenters, for whom breathing is equally important, have it easier. All they need to do is pause between phrases and allow the breath to occur naturally. However, this is easier said than done. Most presenters are so stressed by the pressure of business presentations that they ramble on without pausing—or breathing.

In a recent Power Presentations program, one young woman had such a dread of speaking in front of an audience that she raced through each of the program's morning exercises, anxious to get back to the safety of her seat. Listening to her was, as the saying goes, like trying to take a sip of water from a gushing fire hose. Later in the day, when she finally learned to pause between phrases and stop racing, she was finally able to relax—and even smile. She had given herself what emergency medical technicians give to shock victims: oxygen.

If you're a presenter who experiences the same dread, you don't need Dame Kiri to push your stomach. All you have to do is pause.

Try it—the breath you take will contribute to your own lifetime.

# 48

## The One-Eyed Man
### *Necessity Is the Mother of Invention*

In the sixteenth century, Dutch humanist Erasmus wrote, "In the kingdom of the blind, the one-eyed man is king." Inherent in that statement is that a disadvantage can be an advantage. In the twentieth century, President John F. Kennedy frequently referred to the Chinese symbol for "crisis," which is composed of two characters. One represents danger, and one represents opportunity.

As a presentation coach and writer, a major part of my technique deals with eye contact. But during a lecture tour about presentation skills—which, unsurprisingly, dealt with eye contact—I contracted an eye infection and had to deliver two presentations wearing a distinguished, but nonetheless distracting, patch over one eye. You might think that this challenge impacted my presentation negatively, but it had the opposite effect. The eye patch became my opportunity.

As you've read throughout this section, the overarching principle of the Power Presentations methodology is to consider every presentation as a series of person-to-person conversations. Whenever I present, I role-model that technique by engaging with one person at a time—with every person in every audience. Each time I engage with a different person, I rotate my head and shoulders to face that person directly, but the eye patch caused me to emphasize that technique and to make my rotations more deliberate. In every engagement, every person saw one of my eyes and one eye patch. But in each case, there was no doubt that we had connected. Imagine the power of doing that with both eyes.

I turned my minicrisis into opportunity. Seize the opportunity. Always present person-to-person—and always make eye contact.

# 49

## Bill Clinton's Talking to Me!
### *The Power of Group Dynamics*

A young woman—let's call her Grace—was a participant in a Power Presentations program where she learned the fundamental methodology of treating every presentation as a series of person-to-person conversations. She came up to me after the program to share a relevant personal experience

Grace had been involved in organizing an event for the Graduate School of Education of the University of Pennsylvania, at which former President Bill Clinton was to speak. When Mr. Clinton got to the auditorium, he saw that most of the audience had clustered in the back rows, leaving the front rows vacant. (Perhaps the educators were copying students who sit at the back of the room to avoid be called upon by the teacher.)

Mr. Clinton asked Grace to invite people to move forward, which she did, and then she sat in the front row herself. As his speech proceeded, Grace observed Mr. Clinton's celebrated charismatic style. She was particularly aware of his person-to-person approach and saw that he made eye contact with her quite often. She assumed that he was paying particular attention to her because of her role in organizing the event and because of their exchange about the seating arrangements. But after the speech, Grace spoke with several other people in the audience who reported that they thought that he was paying particular attention to them.

Grace—and all the other people—were experiencing the power of group dynamics. Underlying those dynamics are two powerful forces. The first force is social. When a speaker moves around the audience, engaging one person at a time, and then returns to any one person, that person feels a direct connection with the presenter and a shared connection with fellow audience members.

The other dynamic is the depth-of-field factor in human vision. At a certain distance from the front of the room, audience members cannot see exactly where the presenter's eyes are focused. Think of the time you were in a classroom and the teacher called on someone by gesturing in your direction. Because you couldn't see exactly where the teacher's eyes were focused, you looked around and said, "Who me?"

Those same dynamics are available to you whenever you present— but only when you treat every presentation as a series of person-to-person conversations. If the technique can earn Bill Clinton millions of dollars in speaking fees every year, it can work for you. (F49.1)

# 50

## Liddy Dole and Person-to-Person

### *From Law School to the Republican National Convention*

The Commonwealth Club in Silicon Valley invited me to give a presentation about how to give a presentation. As I always do on such occasions, I spoke about the foundation of delivering every presentation as a series of person-to-person conversations. To demonstrate the technique, I showed a video clip of Elizabeth Dole at the 1996 Republican National Convention. Mrs. Dole, also known as Liddy, delivered a speech in support of her husband, Senator Bob Dole, as the Republican candidate for president. Shortly after Liddy began to speak, she departed from the usual delivery from the dais, picked up a hand microphone, and walked down to the convention floor. From there, she proceeded to walk about, making a series of person-to-person engagements with several people who had been involved in projects with her husband. Her presentation, broadcast on national television, was the hit of the convention.

After my presentation, Bill Peacock, a member of the Commonwealth Club's Silicon Valley Board of Advisors and the master of ceremonies for that event, concluded the evening with an anecdote about Ms. Dole. She was a fellow classmate of Mr. Peacock's at Harvard Law School, and they both were involved in a special project to present a legal brief to a group of sitting judges, one of whom was a member of the U.S. Supreme Court. The law students arrived for their presentations armed with thick binders filled with their research notes, and each of them set the binders on a lectern from which they addressed the panel of judges—except for Liddy Dole. She started from behind the podium for just a few moments and then walked out to speak directly to each judge in a series of person-to-person engagements. Her presentation was so effective, she was declared the winner.

Liddy Dole learned the power of person-to-person conversations at an early age, and she carried that insight throughout her career. It worked for her, and it can work any presenter. It can work for *you*.

# 51

## Fast Talking
### *Fun or Maddening*

An actor named John Moschitta, Jr. once made a career out of talking fast. During the 1980s, Mr. Moschitta appeared in a series of television commercials for Federal Express that promoted the company's speed of service. In each of the commercials, Mr. Moschitta played the role of a busy executive speaking on the phone, barking orders and wheeling and dealing at more than 500 words a minute. The funny, catchy ads made "The FedEx Man" a household name. You can find a clip of his FedEx commercials on YouTube. (F51.1)

However, fast talking can also be maddening for presentation audiences. An accelerated pace often causes a speaker to utter "ums" and "ahs." Speed can also have a negative impact on audiences—particularly for audiences who speak a different language than the presenter. Olivier Fontana, a Frenchman who works at Microsoft Corporation in Redmond, Washington, described his frustration in an email:

*One topic that is a pet peeve of mine is talking speed. I always try to make a conscious effort to slow down when I present—not always successfully—and I also try to explain to native English speakers talking to a non-native audience at full speed that they also need to factor the non-native variable [in]to their speech speed. I received feedback from third parties on how they would sometimes lose more than 50 percent of what was said during these full-speed native speaker presentations. I would also highlight that, to a non-native audience, not making the effort to slow down could be perceived as assuming everyone should understand your language (most often than not, English) perfectly—a perception that could, in some cultures, be perceived as quite arrogant.*

Ironically, there is a very simple solution for fast talking. No, the solution is *not* to slow down. You cannot speak slower or faster. Most people cannot alter their native tempo. I was born and raised in New York City, and I speak fast. I cannot slow down. If I were to try to slow down, it would ... sound ... as if ... my ... battery ... is ... running ... out. Instead, I speak very quickly, but I pause between phrases. Pausing controls my tempo. It can control yours.

Every pause offers many benefits:

- Eliminates *un*words, such as "um" and "ah"
- Enables you to take a breath
- Gives you time to think
- Gives your audience time to absorb

The absorption time is doubly important for people such as Mr. Fontana and other Frenchmen—and Spaniards, Chinese, Japanese, Russians, Germans, and Italians—because when people for whom English is a second language listen to English speech:

- They must translate the English words into their native language.
- They must interpolate the English pronunciations from the perspective of their own language.

Pauses give non-native audiences time to translate and interpolate. The pause also works in the opposite direction. When native English speakers listen to a non-native speak English with a foreign accent, they, too, must interpolate.

In this increasingly globalized world, the solution for language differences is the pause. Think about that: To help your audience comprehend, it's not what you do; it's what you do *not* do.

Pause.

# 52

## Presentation Advice from Titian

### Position, Position, Position

Influential Italian Renaissance artist Titian (1490–1576) painted a portrait of Pope Paul III that established a point-of-view technique used by today's photographers and cinematographers—a technique that every presenter would do well to heed.

In the portrait on display at the Louvre museum in Paris, the point of view of Titian, the painter—and, therefore, the viewer—is looking up at the pope, emphasizing his high social status. Conversely, the pope is looking down at the painter—and the viewer.

The audio guide to the exhibit calls this position the "sociological role" because the angle reinforces the pope's exalted position. To prove the point, the guide goes on to describe the impact of the portrait: "When Titian brought the painting out in the open air for varnishing, passersby bowed down and removed their hats in reverence."

In presentations, your goal is not to create reverence, but empathy with your audience. Empathy occurs when you are at the same eye level as your audience. For maximum impact, therefore, sit when you present—unless you are in a large presentation venue. If the size of your audience or the sight lines of the room challenge your ability to see everyone, stand when you present so that you can look every member of your audience straight in the eye. Eye contact trumps position.

However, standing introduces a new psychological element. In photography and cinematography, when the camera points down at the subject, it is called an "inferior angle," when the camera points up at the subject, it is called a "superior angle," like Titian's portrait of the pope.

Most mission-critical presentations—where a "yes" or "no" hangs in the balance—occur in small groups seated around a conference table. In such settings, presenters often stand and look down at their audience, creating an inferior angle—a negative position for the valued decision makers. So when you present in small groups, sit and engage with every member of your audience at eye level.

They won't remove their hats in reverence, but they will find you empathic. To paraphrase the old adage about real estate, in which location, location, location is paramount; in presentations, position, position, position is paramount.

# 53

## Presentation Advice from Musicians and Athletes
### *The Value of Effortlessness*

Three famous musicians and two athletes share a performance quality that any presenter would do well to emulate. The musicians are jazz pianist Art Tatum, violinist Jascha Heifetz, and dancer Fred Astaire; the two athletes are baseball great Joe DiMaggio and any good trapeze artist. All of them are celebrated for performing their specialties with supreme effortlessness—or, in the idiom of trapeze artists, without a net. The lesson for presenters is to stand up in front of a mission-critical audience and appear supremely confident in describing their businesses. But this is far easier said than done because presenters, unlike musicians and athletes, are *not* performers.

In a *Wall Street Journal* article, Terry Teachout, the newspaper's drama critic, referenced Mr. Tatum's effortlessness in a YouTube video of his 1954 performance of "Yesterdays" and then commented:

*Close your eyes and it sounds as though someone had tossed a string of lit firecrackers into the Steinway. Open them and it looks as though you're watching a court reporter take down the testimony of a witness in a civil suit.*

Mr. Teachout went on to describe Mr. Heifetz's ease: "[He] brought off his stupendous feats of technical wizardry without ever cracking a smile or looking anything other than blasé." (F53.1)

Fred Astaire and Joe DiMaggio were both noted for the consummate grace with which they performed their vigorous physical activities.

How can a presenter achieve that appearance of effortlessness? The answer goes back to the old vaudeville joke about a visitor to New York City seeking directions. The visitor stops a man on the street and asks, "How do I get to Carnegie Hall?" The man replies, "Practice."

The best way to practice your presentation is to Verbalize, the subject of Chapter 13, "Do You Know the Way to Spanish Bay?" Verbalization means rehearsing your presentation aloud—just as you will deliver it in front of an actual audience, and doing it many times over. You can achieve your own effortlessness by implementing a further variation of the presentation equivalent of the real estate credo "Location, location, location" with practice, practice, practice, or, Verbalize, Verbalize, Verbalize.

# 54

## Presentation Advice from Vin Scully

### From Reagan to Barber to Scully

Cable, satellite, and over-the-air television make sports pervasive in our lives. The voices of play-by-play announcers and color commentators fill the airwaves. Most of them are just that: filler—stuffing the soundtrack with meaningless digressions, infantile inanities, vain attempts at jock humor, or, at best, statements of the obvious.

One voice stands out from all the rest: Vin Scully, the radio voice of the Los Angeles Dodgers. Mr. Scully, who has spent more than 60 years as a broadcaster, is widely acknowledged to be the best in the business. The *Wall Street Journal* recognized his talent in a laudatory profile. Mr. Scully defined the secret of his success to his interviewer: "I don't announce," he said. "I have a conversation."

Vin Scully learned his unique style from his mentor, Red Barber—the radio voice of the Brooklyn Dodgers, the predecessor organization of the Los Angeles Dodgers. Mr. Barber, as described on the Radio Hall of Fame web site, was "a down-to-earth man who not only informed, but also entertained with folksy colloquialisms." (F54.1)

An even earlier influence for Mr. Scully had to be Ronald Reagan, whose origins as the Great Communicator go back to the early 1930s. Mr. Reagan was a sports announcer at a radio station in Des Moines, Iowa, where his job was to sit in a studio and describe the play-by-play of Chicago Cubs baseball games from a telegraph ticker tape, as if he were in the ballpark, projecting himself across time and space and, by extension, into the homes of his radio audiences. Then and there, Ronald Reagan learned the art of being conversational.

Reagan to Barber to Scully—a triple play of consummate conversationalists. Make them the role models for the secret to your success as a presenter: Consider every presentation as a series of person-to-person conversations.

# 55

## "Ya' Either Got It or Ya' Ain't"
### *The Fear of Public Speaking Is Universal*

One of the most commonly held *mis*conceptions about public speaking is that good speakers are born that way—meaning, in the lyrics of a song from Stephen Sondheim's classic Broadway musical, *Gypsy*, that "Ya' either got it, or ya' ain't." (F55.1) If any speaker were to accept this false belief, he or she would never be able to change—and presentation coaches would be out of business.

I'm pleased to report that the presentation trade is alive and well, primarily because of the pervasiveness of one of the most common maladies known to humankind: the fear of public speaking. Underlying that fear is another fear: the fear of failure. After all, public speeches and presentations are high-profile events in which the outcome hangs in the balance of success or failure in front of a mission-critical audience. All presenters, whether they are political candidates seeking votes or businesspersons seeking to raise capital or sell a product, face this pivotal juncture. But professional performers also face it because their very livelihood depends on their ability to hold audiences spellbound.

Actor Sir Laurence Olivier, singer Carly Simon, and pianist Glenn Gould have all acknowledged their extreme stage fright. That celebrated list was lengthened by Terry Teachout, the theater critic of the *Wall Street Journal*. In an article on the release of a book about jazz great Benny Goodman's famous 1938 Carnegie Hall concert, Mr. Teachout quoted Mr. Goodman's daughter, who wrote that her father was "always fearful of losing the ability, reputation, and money that he'd gained." (F55.2)

Mr. Teachout went on to cite the apprehensions of choreographer Jerome Robbins and actor–director Orson Wells. Mr. Robbins "left behind a journal in which he set down on numerous occasions his belief that the world would someday realize that 'I'm not talented.'" And Mr. Wells once said, "We need encouragement a lot more than we admit, even to ourselves."

Nevertheless, every one of these famous artists observed the classic Show Business axiom, "The show must go on." In spite of their apprehensions, each of them suited up, and showed up, just as you must do. The next time you experience butterflies in your stomach, know that you are not alone; just make those butterflies fly in formation.

# 56

## How to Eliminate the Fig Leaf

### *A Presentation Lesson from the Military*

When President Obama formally submitted his 2011 budget, he assembled his economic team to join him during his speech. In the official White House photograph, the president is flanked, left to right, by Christina Romer, Chair of the Council of Economic Advisers, Timothy Geithner, Secretary of the Treasury, Peter Orszag, Director of the Office of Management and Budget, and Larry Summers, National Economic Council Director.

**Figure 56.1 "The Fig Leaf" position**

Please note that all the advisors are standing with both their hands clasped below the waist in what is known in the presentation trade as "The Fig Leaf." This position has a number of variations: both hands clasped above the waist, one hand clasped at the midsection, the other hand clasped at the midsection, and one or both hands in the trouser pockets.

All these positions share one common factor: the upper arms are pressed against the sides of the body, and the forearms and hands are held in front of the body. Whenever any presenter, whether a member of the president's staff or you, gets into a stressful situation, such as standing exposed in public, the instinctive response is to protect the midsection where the vital organs are located. This reflex action, also known as "Body Wrap," is an expression of the Fight-or-Flight syndrome and is common to all human beings. In fact, all human beings have been assuming this pose since *before* they were born: Body Wrap is essentially an extension of the fetal position and, therefore, a deeply ingrained habit.

Does this mean that you are forever doomed to strike a defensive pose whenever you appear in public? No. You can eliminate the Body Wrap in many ways, but let's focus on just two. The first is to do what is done in the military: Stand at Parade Rest—the position in which the hands are clasped behind the back and the feet are planted at shoulder width. This brings your shoulders back and your head up, making you appear stable, poised, and confident. However, Parade Rest is applicable only when you arc *not* speaking, such as when posing for an official company photograph or in a group presentation where one person speaks at a time.

The second and most effective alternative is to use your hands and arms to illustrate your words with expressive gestures. These gestures also release the tension of the stress, which caused the Body Wrap in the first place. However, do *not* attempt to script your gestures. Far too many presenters attempt to choreograph their gestures and wind up appearing robotic. Do what comes naturally.

Present with your arms in Body Wrap, and you will appear defensive. Stand at Parade Rest, and you will bring your audience to attention. Gesture and you will hold their attention.

# 57

## *Unwords*
### *Even Barack Obama Says Them*

Every crossword puzzle in the Sunday edition of the *New York Times* has a stated theme. One, called "Roughly Speaking," played out the answers in words that contained either "er" or "um." When spoken, those two sounds are known as fillers or "*un*words" because they have no meaning. Unwords are the bane of any speaker's existence because they make the presenter appear uncertain.

Barack Obama, whose superior speaking skills are acknowledged by detractors and supporters alike, often lapses into *un*words when he departs from his trusty teleprompter and speaks extemporaneously, such as in press conferences. Those lapses drew much spoofing during the early months of his presidency. Many of the spoofs took the form of "slice and dice" videos in which video editors extracted Mr. Obama's "ums" and "'ahs" from the recording of a press conference and strung them together into a tight montage known as a mashup. David Letterman, whose show has a running feature of satirical videos about presidents, did his own mashup of Mr. Obama's *un*words. (F57.1)

*Un*words, as the *Times* crossword puzzle had it, are indeed rough speaking, but consider an opposite point of view. An article in London's *Daily Telegraph* described an academic study in Scotland:

> *Experts at Stirling and Edinburgh universities asked volunteers to listen to a series of sentences, including a number punctuated by "ers" and "ahs."*
>
> *Then they tested how much the listeners could remember, and found that inserting the "ers" had a significant positive effect on how well the subjects recalled what they had heard.*
>
> *Up to an hour after hearing typical sentences, volunteers got 62 percent of words correct where there had been an "er" in the sentence.*

*That compared with 55 percent for similar utterances where there had not been any stumbles. The tests have since been replicated twice and the results are said to be "statistically significant." (F57.2)*

Evidently, something was lost in translation coming across the Atlantic, because *un*words are considered anathema for U.S. presenters. Speakers fret about saying them and search desperately for ways to stop saying them.

Unfortunately, most of the solutions are cast in negative terms, such as, "Don't say 'Um'!" Or, "If you say 'um' you have to pay a quarter!" Negativity doesn't work for nail biters or smokers, nor does it work for presenters. If you tell someone what *not* to do, it will cause that person to do it more often. Telling a presenter not to say "um" will simply produce more "ums."

The simplest and most effective way to eliminate *un*words is to pause between phrases and take a breath. When you take a breath, you cannot make a sound. Try it. Take a deep breath and try to say "um."

See?

Barack Obama did. Realizing the detrimental effects of his *un*words, he took control of his cadence in his extemporaneous press conferences by inserting pauses. In fact, all he did was to leverage the technique that serves him so well in his teleprompter speeches and bring it into his unscripted press conferences.

Speaking from a teleprompter involves reading a script that scrolls by on two transparent teleprompter panels. The separation between the panels causes the speaker—Mr. Obama, in this case—to swing between them; the swings cause him to pause between his phrases, and the pauses inhibit his *un*words. Now, by punctuating his phrases and pausing more frequently, he produces the same rhythmic pattern in his press conferences that is so compelling when he delivers his scripted teleprompter speeches.

Think about that: The way to eliminate *un*words is to breathe.

Yes, you can.

# 58

## To Slip or Not to Slip
### Been There, Done That

In 2010, General Stanley A. McChrystal, the leader of American and NATO forces in Afghanistan, shared his feelings about his Commander-in-Chief with an advisor, who was later quoted in a *Rolling Stone* article:

> *Obama clearly didn't know anything about him, who he was. Here's the guy who's going to run his f°°°ing war, but he didn't seem very engaged. The Boss was pretty disappointed. (F58.1)*

The very same week the article was published, his Commander-in-Chief relieved General McChrystal of his command.

During the same month, Tony Hayward, the CEO of BP, in a statement to the media about his company's disastrous oil spill in the Gulf of Mexico, said, "I'd like my life back." (F58.2)

One month later, BP removed Mr. Hayward from his role as the company's spokesman and, a few weeks after that, from his role as CEO.

In summer 2006, at a rally during his campaign for the Virginia senate seat, Senator George Allen, the favored incumbent, mocked a student of Indian descent as "Macaca." (F58.3) On election day, Senator Allen went down in defeat.

In 2002, during a 100th birthday party for segregationist Senator Strom Thurmond, Senator Trent Lott, the Republican senator from Mississippi and Senate Majority Leader, said:

> *When Strom Thurmond ran for President, we voted for him. We're proud of it. And if the rest of the country had followed our lead, we wouldn't have had all these problems over all these years, either. (F58.4)*

Three weeks of public furor later, Senator Lott resigned his majority post.

Clearly, the World War II slogan "Loose lips sink ships" also applies to politics and business. Politicians lose elections, and businesspeople lose jobs and sink deals as a result of verbal gaffes.

Jeffrey Zaslow wrote an article in the *Wall Street Journal* called "Keeping Your Foot Away from Your Mouth," in which he looked at some famous gaffes by famous people and analyzed the reasons such slips happen:

> *There are gaffes that result from clueless thinking or unfortunate phrasing, and then there are gaffes—such as those with racial or sexual overtones—that can be rooted in our personal belief systems.*

His article went on to note that that the viral influence of the Internet and YouTube have worsened the negative impact of gaffes. "Even if we don't mean it, it can be hard to recover. We've become a culture that is unforgiving when it comes to poor word choice." (F58.5)

Granted, but we can look at verbal errors another way. In business, the high stakes involved in presentations and the resultant pressure are universal givens. Therefore, whenever a presenter makes a mistake, every person in every audience has been there, done that, and so they respond empathically instead of critically. Audiences are forgiving of imperfections that fall into the "clueless thinking or unfortunate phrasing" category—but not for "racial or sexual overtones."

This is not to say that you should intentionally make mistakes to create empathy (although some presenters do this to manipulate their audiences' emotions), but that you should be forgiving of yourself. Many presenters, in striving for unattainable perfection, memorize their presentations. However, memorization (unless you are performing William Shakespeare) is actually counterproductive. If you miss one word, you lose track completely.

Instead, use Verbalization—the efficient way to practice that you've read about several times in this book. The reason for the repetition here is that Verbalization is one of the most powerful tools available to presenters—yet one of the most underutilized.

But even Verbalization does not guarantee perfection. When, not if, but *when* you slip in a presentation—or in any situation in life—remember the words of eighteenth-century British poet Alexander Pope: "To err is human, to forgive divine."

# 59

## The Free Throw
### A Presentation Lesson from Basketball

"It takes me a couple of minutes to settle down" is one of the most common utterances presenters make, and they make it in reference to the jolt of adrenaline that every presenter—novice or veteran—experiences at the start of every presentation. The jolt produces the Fight or Flight reaction to stress, which, in turn, causes the presenter's eyes to dart around the room—regardless of the size of the room or the size of the audience—in search of escape routes; the darting then makes that person appear furtive to the audience. If you'll pardon the play on words, in the blink of an eye, the audience gets an immediate negative impression of the presenter, and there is never a second chance to make a first impression.

Even if the veteran ultimately settles down, the die is cast. Both veteran and novice would do well to defuse the jolt and start strong. The veteran will not have to wait the few minutes, and the novice can get started on the right foot.

A lesson in how to control this pivotal first moment comes from how basketball players shoot free throws. Think about their process. Even though these well-trained athletes have practiced this shot countless times, they go through the same brief ritual each time. They step to the foul line, look at the basket, bounce the ball a few times, take a deep breath or two, look at the basket again, and then shoot. Whether or not they make the first shot, they go through the same ritual for the next shot. In each case, the intent of the ritual is to stabilize the player *before* the shot.

To stabilize yourself *before* your presentation, try this ritual. As you head up to the front of the room, have the first few phrases of your presentation set in your mind—phrases that you have practiced enough to be familiar with them. When you get to the front, the jolt of adrenaline will force your eyes to sweep the room in search of flight paths.

You cannot stop this involuntary reaction, so let the sweep happen. But let your eyes move around the room with an appropriate action: your welcoming remarks. "Good morning. Thank you all for your time. I appreciate the opportunity to speak to you." By combining your sweep with your welcome, your movement will appear natural instead of furtive.

Then, having swept the room, stop. Turn to the other side of the room and find a person in the audience a considerable distance away from the last person at the end of your sweep. This wide swing is helpful for two very good reasons. One is to give you a fuller pause to think—thinking is always a good idea—and the other is to give you a moment to settle in, just as the basketball player settles in at the foul line.

And just as the basketball player focuses on the basket, focus on that new person. Take a breath. Then deliver the first well-practiced phrase of your presentation to that person. Then turn to another person and deliver your well-practiced second phrase to that person. Continue around the room to deliver additional phrases to other people.

These simple steps will nullify the negative effects of your adrenaline rush and make your first impression positive.

Nothing but net.

# 60

## 10 Tips for 30 Seconds
### *Help for Job Seekers*

In response to the difficulties caused by the economic downturn, CNN provided an excellent public service to help unemployed people find work. The cable news channel gave job seekers 30 seconds on air to make a pitch to prospective employers in the viewing audience. Below you'll see ten helpful pointers for the specialized circumstances of a television appearance, but you'll also see that many of them are applicable to an in-person, face-to-face interview, and even to a presentation.

1. Prepare thoroughly. Plan your content carefully. Rehearse repeatedly. Time yourself. Rehearse *aloud*. You will recognize this method as Verbalization. Rehearse until you are very familiar with your content, but *not* to the point of memorization.

2. Get plenty of sleep the night before the television studio session. It makes a big difference.

3. Sip warm water just before you start. Avoid coffee, milk, and carbonated beverages.

4. Sit forward, with your feet planted squarely on the ground; this will make you appear alert and poised.

5. Sit on the coattail of your jacket; this will keep your collar from pulling away from your neck.

6. Make eye contact with the camera lens, but don't stare. Allow yourself to blink.

7. Smile. No explanation needed.

8. Say your name.

9. Say "you," referring to your prospective employer.

10. Offer benefits to your prospective employer. Say what your skills can provide to a company that hires you. In other words, give them WIIFYs (What's In It For You).

# 61

## You Are What You Eat
### *Ten Tips about Food and Drink in Presentations*

"You are what you eat," a phrase that has become common in today's lexicon, actually came into being in the nineteenth century. In 1825, Jean Anthelme Brillat-Savarin, a French lawyer, magistrate, and politician, published *The Physiology of Taste: Or, Meditations on Transcendental Gastronomy*, in which he wrote, "Tell me what you eat, and I will tell you what you are." (F61.1)

Over time, Monsieur Brillat-Savarin's treatise on cooking and eating has become a bible for foodies, and his phrase, in its shorter form, has become a slogan for dieticians.

Brillat-Savarin's modern counterpart, food guru Michael Pollan, the author of the bestselling *Omnivore's Dilemma, Food Rules* and *In Defense of Food: An Eater's Manifesto*, has created a slogan with similar advice: "Eat food. Not too much. Mostly plants." (F61.2)

Just as food intake influences our physical beings, it also influences our presentations. Here are ten tips for you to keep in mind when you present:

1. Eat lightly. Follow Michael Pollan's advice: Have a small salad or clear soup or broth, at most.
2. Avoid nuts or crackers that stick in your teeth.
3. Sip—do not gulp—water to moisten your mouth and throat.
4. Take a nature break just before you present.
5. Use sport-top bottles to avoid spilling and sloshing.
6. Avoid coffee, tea, Red Bull, alcohol, or other stimulants.
7. If you need a stimulant, use the natural sugars in fruit juices.
8. Avoid dairy products, as they coat the throat with a mucosal film.
9. Avoid carbonated beverages.
10. If you have a cold or cough, use honey or medicinal teabags, such as Throat Coat.

# Section
## IV

Q&A: Handling Tough Questions

# 62

## Speed Kills in Q&A
### *The Vanishing Art of Listening*

One of the most important qualities for success in business is the very quality that impedes the effective handling of tough questions: rapid response time. Every businessperson is expected to react quickly to problems and to come up with prompt solutions. But in responding to tough questions, speed can kill.

Tough questions are a part of the territory in business and, after the Great Recession, the terrain is rougher than ever. In every facet of life, people are in search of answers to their problems, so their questions are loaded with emotion. If a responder answers too rapidly and with equal emotion, be it defensive or contentious, the battle is joined and the exchange heads rapidly downhill—a lose-lose engagement.

In preparing for tough questions, a results-driven mindset often involves an approach know as "Rude Q&A," in which a list of anticipated challenging questions is assembled and then matched with a list of appropriate answers. This approach has a small flaw: People don't ask questions as written; most questions come out in long, rambling, and convoluted sentences. This causes the responder to scramble for the right answer, at best, or the wrong answer, at worst.

The solution is to slam on the brakes and *not* think of the answer. Stay in the moment. As the question is being asked, listen carefully and identify the central issue embedded within the convolution.

What a concept: Listen! Listening is becoming a lost practice in our culture. For those people who still retain a semblance of politeness, listening has become a matter of waiting for one's turn to speak. And for those sadly increasing numbers of people who no longer bother to listen, the practice has converted to talking past the next person.

For the results-driven businessperson, listening will feel counterintuitive and will be difficult to do, but it is absolutely vital. Failure to observe this simple rule can result in failure of the answer, the presentation, the meeting, or the entire business proposition.

When you are asked a question, follow the advice of the old adage frequently attributed to Epictetus, a first-century Roman philosopher, "We have two ears and one mouth, and we should use them in that proportion."

The full solution is called Active Listening, a subject that brings up more than eight million entries in an Internet search, but the essence of the skill can be summed up in one sentence: When you are being asked a question, use the powers of your mind to focus on the essence of the question, *not* on the answer.

Listen.

# 63

## A Lesson in Listening from Barack Obama
### *How to Handle Multiple Questions*

Listening, the social skill that is becoming extinct in the twenty-first century, is rapidly being replaced by rudeness. This behavior, which is merely annoying in social circles, can be destructive in the more mission-critical circumstances of business and politics.

One of the most common instances of not listening comes in response to multiple questions. Because such queries are usually a rambling set of unrelated issues, it is difficult for any presenter to remember all the diverse parts. Two common responses are to answer only one of the questions and then to ignore the second question by moving on to another questioner, or to answer the first question and then to turn back to the questioner and say, "What was your other question?" Both of those responses create the perception of not listening. Not listening leads to failure.

One way to handle multiple questions is to pick only one, answer it, and then turn back to the questioner and say, "You had another question." Note the difference between that statement and "What was your other question?" which is a question about the question. This technique will result in one of two responses from the questioner. That person will either tell you what the other question was or say, "That's okay, you covered it." Both results take you off the hook.

Barack Obama used another technique for handling multiple questions. In a White House press conference, Jeff Zeleny of the *New York Times* asked him, "During these first 100 days, what has surprised you the most about this office, enchanted you the most about serving in this office, humbled you the most, and troubled you the most?"

The president immediately reached into his coat pocket, pulled out a pen, and said, "Let me write this down," producing a wave of laughter from the other reporters in the East Room of the White House.

As Mr. Obama began writing, Mr. Zeleny began to restate his question, "Surprised ... troubled ..."

Mr. Obama said, "I've got—what was the first one?"

Mr. Zeleny repeated, "Surprised ..."

Mr. Obama repeated, "Surprised ..."

Mr. Zeleny repeated, "Troubled ..."

Mr. Obama repeated, "Troubled ..."

Mr. Zeleny repeated, "Enchanted ..."

The president said, "Enchanted," then smiled and added, "Nice," evoking more laughter. (F63.1)

Barack Obama had carefully listened to the question, confirmed that he had listened by restating what he heard, and then reconfirmed by writing what he heard, sending a clear message of his attentiveness.

Compare his approach to the usual evasive response from most politicians. We have learned to tolerate, if not endure, nonanswers from politicians, but no man or woman can get away with evasiveness in business. However, business presenters often give the *appearance* of evasiveness because they handle multiple questions poorly. They do so not to evade, but because they don't address a convoluted question fully. Nevertheless, the net effect is the same: not listening.

Take a lesson from the president. Next time someone asks you a set of long, rambling multiple questions about your presentation, reach for your pen, start writing, and confirm what you heard. Or simply pick one of the questions, answer it, and then say, "You had another question." Just as the *Times'* Jeff Zeleny did, your questioner will repeat the question.

Use your brain to think about the question and the answer, and leave the remembering to the rambler.

# 64

## If I Could Tell Jon Stewart...
### *Talk Shows Include Listening*

After a long, intensive career in broadcasting, I went cold turkey on the medium and stopped watching television, except for news, football, and a few select programs. One of the select of the select, the only series on my DVR, is *The Daily Show*. Its appeal:

- **Format.** Pure television, not televised radio
- **Intelligence.** Adult, not talking down to the audience
- **Expression.** Both sacred and profane, not bland pap
- **Timeliness.** Current, not designed for reruns
- **Point of view.** Innovative, not imitative

And the greatest appeal is Jon Stewart, the star and spirit behind the series, whose talent (and, frequently, humor) is over the top. I'm addicted to the show.

Yet if I could offer one piece of advice to Mr. Stewart, it would be to do more listening to his guests during interviews and less interrupting. Jon Stewart's illustrious predecessor, Johnny Carson, the king of late-night television, got as many laughs per minute as does Mr. Stewart, but Mr. Carson drew more out of his guests. The king's approach: listening and reacting. His reaction, more often than not, was a silent mug.

Jon Stewart's rubbery face can mug with the best of them. His repertory of facial expressions is as broad as that of Red Skelton or Stan Laurel and Oliver Hardy. If you think about it, the comic talent of those classic clowns was as much in what they did as in what they said. Silence is golden.

Listen and react. It worked for Johnny Carson, it can work for Jon Stewart, and it can work for you.

# 65

## What Keeps You Up at Night?

### *How to Handle the Most Frequently Asked Questions*

"What keeps you up at night?" and "What is your greatest weakness?" are perhaps the two most frequently asked questions (FAQs) in business—the first in presentations and the second in interviews. Yet both questions, by their frequent recurrence, are traps for glib answers that could derail the person who provides the answers.

Joann Lublin's career column in the *Wall Street Journal* offered advice about how to handle the interview FAQ about weakness, starting with what *not* to say. She provided a long list of common glib answers, ranging from "I am a perfectionist" to "I am a workaholic"—all of which offer a strength instead of a weakness and, therefore, appear evasive. Ms. Lublin recommended better, more candid answers, such as having a "tendency to make decisions too fast." But then she concluded with the most important piece of advice: that any answer to such a question should "cover your corrective steps." (F65.1)

This same advice is also applicable to the "What keeps you up at night?" question in presentations. That question has become ritual in every type of presentation and every type of business. It is phrased in those exact words. Not "What problems do you foresee?" Not "What can go wrong?" Not "What are your threats?" But "What keeps you up at night?"

What should you say in response?

What *not* to say in reply to this universal question is to make a joke about newborn babies, neighbors' dogs, air conditioners or the like. Everyone has heard every variation on that lame theme. What *to* say must be purely candid—a direct answer to a direct question. In business, evasion is not an option. With almost daily revelations of public corruption that are met with denial, evasion, or blaming others, transparency has become more important than ever.

Be frank. Tell your questioner what keeps you up at night, but then immediately add what actions you are taking to correct those issues. "What keeps me up at night is _____, and what I'm doing about it is _____."

Accountability is all.

# 66

## Spin versus Topspin
### *The Political World versus the Business World*

In the competitive world of politics, jockeying for position is often expressed by **spinning,** the dark art of attempting to influence public perception in one's favor or against that of the opposition. Spinning can be as harmless as gilding the lily, or it can escalate to distortion or even to outright deception; however, all the points along that scale are of dubious integrity.

One of the best examples of political spin is the 1998 film *Wag the Dog*, in which a U.S. president is accused of a scandalous liaison. To limit his damage, the president calls in a Washington spin doctor, played by Robert DeNiro, who proceeds to retain a Hollywood producer, played by Dustin Hoffman, and together they concoct a fictional war in the Balkans.

In this scenario, spin could more accurately be called "slant," for the tactic diverts attention away from the main issue. Spin is akin to the sleight-of-hand magicians use to misdirect audiences. This is not to say that a politician, a businessperson, a representative of an organization, or anyone in *any* competitive walk of life—including *you*—should not do everything you can to defend your own cause and position it in a favorable light. However, before you do so, you must address the issue directly; only then can you go on to counterbalance the negativity by *adding* your own message.

This additive instead of digressive approach is called **Topspin,** a subject covered in detail in *In the Line of Fire*. Topspin is a tennis term that refers to a power stroke that causes the ball to bounce sharply and give a player a winning advantage. In presentations, Topspin is a positive statement or restatement of a key message that gives a speaker a winning advantage.

But just as tennis players must first meet the ball before applying Topspin, presenters must first address the central issue directly before adding their own message. They must earn the right to state their case. Politicians rarely address issues; businesspersons must always do so. In business, accountability trumps messaging.

For instance, if a salesperson were to be asked by a customer, "Why do you charge so much more for your product than your competition does?" the salesperson could respond, "The reason we sell our product at that price point is because we provide you with a service guarantee that extends the life of the product. When you buy our product, you get more for your money."

Notice that the response doesn't deny the price point, nor does it agree that the price is high. Thus, the salesperson acknowledges the negativity in the question without any evasion, admission, or contention. Then, having addressed the issue directly, the response continues to Topspin with a call to action ("When you buy our product") and a benefit ("you get more for your money").

Topspin is a world apart from spin.

# 67

## When Did You Stop Beating Your Wife?
### How to Handle False Assumption Questions

Sooner or later, every human being is confronted with tough questions. One of the toughest and most common is the infamous loaded question "When did you stop beating your wife?" which implies that you have indeed been beating your wife. How do you answer without agreeing with the implication? How do you *not* answer without appearing evasive?

Courtroom dramas often include a scene in which an antagonistic prosecutor points his finger at a defendant and asks accusingly, "Why did you kill your partner?" implying that the person—who has pleaded *not* guilty—did kill the partner. Or "What did you do with the gun?" implying that the person did possess the murder weapon.

This is known as a false assumption, and the only way handle such a question is to apply the noted antidrug slogan championed by Nancy Reagan: Just say, "No!"

The defendant should say, "I did not kill my partner." The businessperson should deny the false assumption. And if anyone ever asks you when you stopped beating your wife, simply rebut the fallacy by saying, "I never started."

Former Secretary of State Colin Powell is an excellent speaker and a master at handling tough questions. One of his most challenging Q&A sessions came at a press conference he held shortly after the start of the Iraq war. A Lebanese journalist asked him:

*Mr. Secretary, a lot of fears have been made about who is next. And some people believed to be close with the administration said that the regimes backing Cairo and in Saudi Arabia should be nervous right now. How do you address that point? And does the U.S. have a plan to spread a set of values at gunpoint, in your view, at this point? (F67.1)*

"... a plan to spread a set of values at gunpoint ...." This question accused the United States of acting as a villainous bully, and Mr. Powell could not give credence to this charge. When he retook the floor, he immediately countered the accusation: "No, of course not."

Neither Colin Powell, nor you, nor *any* presenter is under *any* obligation to respond to a wrongful accusation in *any* other way than with a complete refutation. If you are attacked with a question that contains or implies an inaccuracy, do as Mr. Powell did. Say, "No."

Or do as Barack Obama did in his first press conference that took place less than a month after he took office. With the financial crisis weighing heavily on the nation, the media honeymoon of his election victory and inauguration ceremonies ended abruptly: 3 of the 13 questions asked by the reporters in that session were false assumptions.

The first was from Associated Press reporter Jennifer Loven, who began by referring to Mr. Obama's statement earlier that same day that the economic crisis might be irreversible, and then asked him, "Do you think that you risk losing some credibility or even talking down the economy by using dire language like that?"

Mr. Obama's first four words were, "No, no, no, no."

Caren Bohan of Reuters then asked, "Did you underestimate how hard it would be to change the way Washington worked?"

The president replied, "I don't think I underestimated it. I don't think the American people underestimated it."

Chip Reid of CBS News asked the third false assumption question: "You talked about that if your plan works the way you want it to work, it's going to increase consumer spending. But isn't consumer spending, or overspending, how we got into this mess? And if people get money back into their pockets, do you not want them saving it or paying down debt first, before they start spending money into the economy?"

Mr. Obama said "no" again. "Well, first of all, I don't think it's accurate to say that consumer spending got us into this mess. What got us into this mess initially were banks taking exorbitant, wild risks with other people's monies, based on shaky assets." (F67.2)

In each case, the president demonstrated his trademark cool demeanor, contradicting his interrogator and then moving on to correct the fallacy by stating his own position on the given issue.

The next time someone asks you a false assumption question, follow Colin Powell and Barack Obama's examples and take Nancy Reagan's advice: Just say, "No!"

# 68

## Madoff and Cramer Plead Guilty
### How to Respond When Guilty as Charged

In an ironic coincidence, two prominent public figures pleaded guilty on the same day in March 2009: Bernie Madoff to a judge, and Jim Cramer to Jon Stewart. Mr. Madoff, caught red-handed on 11 felony counts of swindling stocks, had to confess; Mr. Cramer, caught red-handed of hyping stocks, did not. Hyperbole is not a crime—yet. But if Mr. Cramer's admissions were an effort to tell the public that he intended no malice, he could have done so more positively. Instead, he withered in the line of Mr. Stewart's fire.

Jim Cramer certainly had the wherewithal to stand up for himself in a contentious exchange, such as his appearance on *The Daily Show* (F68.1) was sure to be. As a Harvard-trained attorney (including a stint as a research assistant to the famously contentious Alan Dershowitz) and as a seasoned television professional, Mr. Cramer surely knew a thing or two about handling tough questions. Moreover, Mr. Stewart's questions were not surprising—he had spent the three prior episodes that week trashing Mr. Cramer and CNBC. *USA Today* (F68.2) had touted Mr. Cramer's appearance with a banner headline, and *Business Week* (F68.3) had called it the "week-long match of the century." Mr. Stewart himself previewed the encounter as a "battle between a man who makes people laugh for a living and whatever people think I do."

Yet when Jim Cramer took his place in the hot seat, he capitulated completely. As Jon Stewart repeatedly berated him, Mr. Cramer repeatedly agreed, saying, "Okay," or "You're right," or offering *shoulda'*, *woulda'*, *coulda'* replies. And when not agreeing with Mr. Stewart's charges, Mr. Cramer shifted the blame to "the regulators" or to the "lying CEOs."

Worse still was Mr. Cramer's presentation. On his own CNBC show, *Mad Money*, he usually stomps around the set shouting loudly and waving his arms wildly. Seated across from Mr. Stewart, Mr. Cramer spent most of his time docilely nodding his head in agreement or shrugging his shoulders, his arms splayed open in submission.

Did Jim Cramer prepare himself, as anyone with a modicum of experience with tough Q&A sessions would? Did he anticipate a list of the worst-case questions? In fact, before Mr. Cramer came onstage, Mr. Stewart performed his own mock preparation session: responding to shouted questions from off-screen voices while seated against a bare brick wall, under the glare of a naked light bulb.

What could Jim Cramer have been thinking? If he wanted to make amends to the public he had misled, he could have done so on his own turf, to his own unseen and unheard audience. Instead, he went into the lion's den and submitted to a scathing dressing-down from Mr. Stewart.

Even Jon Stewart was perplexed. In his closing remarks, he turned to the camera and said, "I hope that was as uncomfortable to watch as it was to do."

We all make missteps or mistakes at some point in our careers. If those mistakes are similar to those of Messrs. Madoff and Cramer, they are matters of accountability and are beyond repair. If they are honest mistakes, here is how to handle them: Take a lesson from Jon Stewart's mock grilling. Prepare a list of tough questions about your blunder and have a colleague or colleagues fire them at you in a practice session. In your answers, be completely honest and admit your responsibility, but then promptly go beyond your admission and offer what corrections you have made or will make. Give more weight to your corrections than to your admissions. Repeat the mock grilling until you become fluent and succinct. By the time you step into the real Q&A session, unlike Jim Cramer, *you* will present a positive image.

# 69

## Tell Me the Time, Not How to Build a Clock

### Keep Your Answers Short

"Brevity is the soul of wit," said Polonius, the sage royal advisor in *Hamlet,* in response to the king's request for his opinion. William Shakespeare had his seventeenth-century character use *wit* to mean intelligence instead of its current usage to mean clever humor. But just as the definition of wit has shifted over time, so has the definition of brevity. Far too many presentations—and even more conversations—go on and on and on, warranting the impatient accusation, "Tell me the time, not how to build a clock!"

We have become accustomed to politicians who go into windbag mode at the drop of a hat. The most egregious example is in televised congressional hearings, where even the questions posed by senators or representatives are interminable. Over time, we have become reluctantly tolerant of verbosity in politics, but we never cut anyone such slack in business.

In presentations, nowhere is verbosity more frequently perpetrated than in answers to questions. All too often, presenters introduce new material in their answers or rehash their original material ad nauseam. They fail to understand that the primary purpose of opening the floor to questions is to clarify the content in the presentation. Audiences are less interested in comprehensive and exhaustive responses to their questions than they are in seeing how a presenter handles him- or herself in the line of fire.

The best expression of this view comes from David Bellet, the founder of Crown Advisors International, one of Wall Street's most successful investment firms. David, who was an early backer of Hewlett-Packard, Sony, and Intel, is retired now, but whenever he considered an investment, he often made challenging questions a standard part of his due diligence.

> *When I ask questions, I don't really have to have the full answer because I can't know the subject as well as the presenter. What I look for is whether the presenter has thought about the question, been candid, thorough, and direct; and how the presenter handles himself or herself under stress. (F69.1)*

Follow David Bellet's advice: Make your answers to your audience's questions brief. When someone asks you the time, just tell them the time. Make your presentations brief, too. And while you're at it, make your conversations brief. Conversation is a form of communication, only less formal than presentations. The social windbag is as much of—if not more than—a bore as the political windbag. We can always change the channel on the politician; escaping the conversational chatterbox is more difficult.

Be succinct in your presentations and conversations, and leave the clock making to Timex.

# 70

## Presentation Advice from Jerry Rice
### *Grasp the Question before You Answer*

Jerry Rice played for the San Francisco 49ers football team from 1985 to 2000 at the wide receiver position. He played so well that he set virtually every significant record for receivers. Some of the more notable career records include receptions (1,549), receiving yards (22,895), most 1,000-yard receiving seasons (14), total touchdowns (208), and combined net yards (23,546). (F70.1)

What makes for a successful wide receiver is a statistic called **Yards After Catch**, or its acronym, *YAC*. The statistic refers to a play in which a receiver catches a pass for a gain of yards and then runs for additional yards. Superior receivers, such as Mr. Rice, strive for long YACs. The not-so-superior receivers, in their desire to become superior receivers, often take their eyes off the ball and start to run before they catch the ball. They then fail to make the yards or the catch. The play fails.

The analogy to Q&A sessions applies here. In their desire to succeed, presenters often start to provide an answer before they fully understand the question. If the answer doesn't match the question, the answer—and the entire presentation—fails. The missing link in this equation is listening, a social skill that is rapidly becoming extinct in the twenty-first century.

You've read several examples of the importance of listening in this section; for now, let Jerry Rice's skill reinforce the importance of listening before you answer. Do not take a single step into your answer until your hands are clutching the ball—until you fully grasp the true meaning of the question.

# 71

## Politicians and Spin
### *Putting Lipstick on a Pig*

In the 2010 election, Richard Blumenthal, the attorney general of Connecticut who ran for Christopher Dodd's Democratic seat in the Senate, and Rand Paul, who ran for the Republican Senatorial seat in Kentucky, found themselves having to explain controversial statements they had made during the campaign—Mr. Blumenthal on the subject of whether he had seen active duty in Vietnam, and Mr. Paul on whether he would support the Civil Rights Act of 1964. Each man's original statement had raised a firestorm in the media and on the Web, and each man had to make new statements to clarify his position.

In politics, this backtracking is known as spin, or "putting lipstick on a pig." When the spin doesn't cover the original tracks, even the spinners' supporters look unkindly on the tactic.

- Democratic National Committee Chairman Tim Kaine, appearing on ABC's *This Week,* said of Mr. Blumenthal's controversy, "Those statements were wrong, period. They were wrong and it was very important for him to acknowledge that and clear that up." (F71.1)

- The *New York Times*' Republican columnist, Ross Douthat, characterized Mr. Paul's explanation as "conspicuously avoiding saying that he would have voted for the bill that outlawed segregation. By the weekend (and under duress), he finally said it. But the tapdancing route he took to get there was offensive, tone deaf, and politically crazy." (F71.2)

As we the people have so painfully come to expect, spin does not clarify. At best, it digresses; at worst, it obfuscates. In business, spin is not an option. But politicians spin so often the public has come to tolerate it.

The most egregious example of political spin I have ever seen arrived in my mailbox via an email blast that indicated that it had been forwarded many times. If you receive as many such missives as I do, you are probably as dubious of its validity as I am. The story is very likely apocryphal, but I am taking the liberty of sharing the text with you to demonstrate just how far—and how creative—politicians will go to alter facts. (I have edited the names from the original email to avoid identifying the alleged perpetrator of the spin.)

The email described a professional genealogy researcher who had come across historic evidence that a sitting member of the U.S. Senate had a distant relative who was a horse thief and a train robber during the latter part of the nineteenth century. The relative had an interesting police record: He was arrested, sent to jail, and then escaped, but was ultimately caught and hanged. The researcher wrote to the senator inquiring about this relative. The senator's office staff replied:

> *He was a famous cowboy. His business empire grew to include acquisition of valuable equestrian assets and intimate dealings with the railroad. Beginning in 1883, he devoted several years of his life to government service, finally taking leave to resume his dealings with the railroad. In 1887, he was a key player in a vital investigation run by the renowned Pinkerton Detective Agency. In 1889, he passed away during an important civic function held in his honor when the platform upon which he was standing collapsed.*

That is a shade of lipstick that would make Revlon blush.

# 72

## Murder Boards
### *How Elena Kagan Prepared for Tough Questions*

Before the full Senate approved Elena Kagan, President Obama's second nominee for the Supreme Court, the Senate Judiciary Committee put her through a series of confirmation hearings. Just as Mr. Obama's first nominee, Sonia Sotomayor, and both of President George W. Bush's nominees, John G. Roberts and Samuel Alito, and all the previous nominees of all the previous presidents had endured, Ms. Kagan was grilled mercilessly by the senators, particularly those of the opposition. All's fair in politics, and the party out of power wants to do everything it can to make the sitting president—and that president's choices—look bad.

In preparation for the grilling, Ms. Kagan spent long hours in mock sessions called "Murder Boards." This intense practice process, which includes everything from re-creating the setting in the Senate chamber to anticipating the worst-case questions from the senators, was described in a post on realclearpolitics.com by Julie Hirschfeld Davis. One particular item in the article that deserves attention came from Rachel Brand, an attorney who helped prepare Justices Roberts and Alito for their confirmation hearings. Ms. Brand said that the purpose of the Murder Boards "is to ask those hard questions in the nastiest conceivable way, over and over and over." (F72.1)

The triple iteration of *over* is the operative point. In earlier chapters, you read about Verbalization—the process of rehearsing your presentation aloud as you would to an actual audience. That same practice is just as—if not more—important in handling tough questions. It might seem sufficient to list the anticipated challenging questions and to craft an answer for each of them, but that is not enough. It is far more effective to have someone fire those questions at you and to speak your answers aloud. And you must do it over and over and over. The dynamics of the repeated interchanges in practice will make your responses in real time crisp and assertive.

The CEO of a Silicon Valley company who had taken the Power Presentations program in preparation for his IPO road show decided to prepare for his subsequent quarterly analysts' call by writing the anticipated tough questions on flash cards and Verbalizing his answer to each card. To his dismay, during the actual call, he found his responses halting. He called me for a brush-up, and I told him that the flash cards were not a substitute—even in mock practice—for having a human voice fire the questions.

The Murder Boards for Ms. Kagan did it right. According to the article, the questions fired at her came from "About 20 members of President Barack Obama's team ... Kagan's pals from academia as well as White House and Justice Department lawyers." They made the mock practice more real.

In preparation for your next Q&A session, have a member or members of your team fire tough questions at you, and Verbalize your answers to them over and over and over.

Think of the repetition as volleying to perfect your tennis game.

# 73

## Ms. Kagan Regrets
### Nonanswers to Tough Questions

Cole Porter's 1934 song "Miss Otis Regrets," a wry blues tale about a society lady indisposed to answer questions, had its modern variation during the Senate Judiciary Committee hearings on President Obama's nominee for the Supreme Court, Elena Kagan.

In the previous chapter, you read how thoroughly the president's staff prepared Ms. Kagan for the hearings by subjecting her to Murder Boards—intense practice sessions in which tough questions were fired at her repeatedly and to which she gave her answers repeatedly. Apparently, part of the preparation also included not answering some questions.

Jon Stewart seized on this strategy in his coverage of the hearings on *The Daily Show,* first by setting it up:

*Perhaps this year will be Elena Kagan's chance to demonstrate the proper manner in which to answer committee questions in a forthright, nonevasive, honest, judicially transparent way, so that we may, as a nation, finally have the Supreme Court confirmation conversation that we deserve.*

Mr. Stewart then followed his lead-in with quick cuts of about half a dozen video sound bites from the hearing in which Ms. Kagan refused to comment or said that a comment would not be appropriate. The juxtaposition told the tale. (F73.1)

Linda Greenhouse also seized on the refusals in the *New York Times.* Ms. Greenhouse, who won a Pulitzer Prize in 1998 for her coverage of the Supreme Court and teaches at Yale Law School, wrote, "A hearing like this represents a lost opportunity for the public to actually learn something about how judges think about what the Constitution means." (F73.2)

Because of the highly polarized political aspects of such hearings and the extremely fine points of constitutional law, candidates for the Supreme Court can invoke caution or appropriateness in not answering; because of the public's low expectations of integrity in the political world, politicians often get away with ducking tough questions.

You do not have that option. In business, you must answer every question asked of you.

No ifs, ands, or buts.

# Section
# V

Integration: Putting It All Together

# 74

## The Elephant

### *The Whole Is Greater Than the Sum of the Parts*

In 1873, John Godfrey Saxe, an American poet, published a poem based on an ancient Indian fable about six blind men who were asked to describe an elephant by touch. One man said it was a wall, another a spear, another a snake, another a tree, another a fan, and the last man a rope. The final stanza of the poem concludes:

> And so these men of Indostan
>
> Disputed loud and long,
>
> Each in his own opinion
>
> Exceeding stiff and strong,
>
> Though each was partly in the right,
>
> And all were in the wrong! (F74.1)

The point of both the poem and the fable is to demonstrate the importance of seeing objects—as well as objectives—from an overarching view instead of just as component parts; to see the forest, not just the trees.

Contextual perception also applies to presentations. Conventionally, people in business view a presentation as the individual parts of an elephant. One person describes it as the story, another as the slides, another as the delivery, and yet another as the handling of tough questions.

However, a well-told story can be ruined by a slide show that resembles a doctoral dissertation on quantum physics, or by a presenter stricken by the fear of public speaking, or by a zinger question from the audience.

The presenter must manage every one of these elements. More important, the presenter must integrate every one of these elements with each of the other elements, or any one of them can backfire and ruin the entire presentation.

The presentation is the elephant.

# 75

## Presentation Graphics Meet Linguistics
### *Symmetry in Graphics Design*

Matt Vasey, the Director of the American Distribution Channel at Microsoft Corporation, was a participant in a Power Presentations program held at Microsoft's Redmond, Washington, campus. During the session on graphics design, one of his colleagues showed a bullet slide arranged in the format shown in Figure 75.1. Matt gave his commentary about the content and then concluded, "I'm not crazy about that glottal stop."

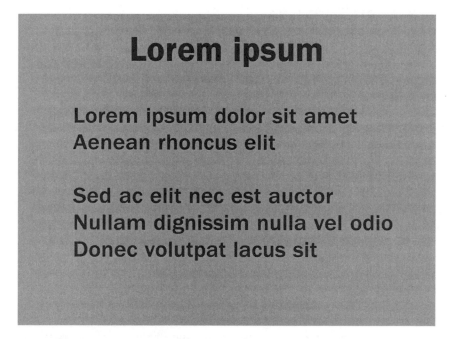

**Figure 75.1   Unattractive spacing in bulleted lines**

His words brought me up short.

A glottal stop is an esoteric phonetics term that refers to an action of the vocal cords snapping shut over the glottis, the space between the cords, during speech. The action produces a sharp, unattractive sound, and Matt was clearly referring to the unattractive spacing in the bullet lines.

I asked, "How do you know about glottal stops?"

Matt replied, "I took a linguistics course in college."

While Matt was using the term analogously to make a point about design (bullet lines should be spaced evenly), he was also inadvertently making another point about the relationship between graphics design and narration.

Unfortunately, in common presentation practice, these elements are often treated as two distinct entities. This separation results in dissociation between what the presenter shows and what the presenter says. The split forces the audience to stop listening while they try to understand the graphic. As a result, the disruption to the presentation is even greater than irregular spacing on a slide or the sound of a glottal stop.

You can bring your design and your words together in a very simple way. Whenever you display a slide, use your narration to help your audience understand what they are seeing. Use a Title *Plus* to describe your slide. A Title *Plus* is a succinct statement that captures the entire content of the slide.

Design each slide with a single-line title that conveys the main point. Each slide also has additional material below the title—bullets, graphs, icons, pictures, tables; that is the *plus*. Every time you click to a new slide, tell your audience what you've shown them using the Title *Plus*:

- "Here you see five years of annual revenues."
- "This table compares our product to all others."
- "These are the many benefits of our product."

State the Title *Plus* the instant the new slide appears, or your audience will stop listening and try to understand what they are seeing. After the Title *Plus*, you can go on to discuss your slide in greater detail and add value.

Synchronize your graphics with your linguistics and you with your audience.

# 76

## One Presentation, Multiple Audiences
### *12 Presenters, 12 Stories, 1 Set of Slides*

The foundation of the Power Presentations methodology is that the presenter is the focus of the presentation, not the slides. But this concept often runs into objections that usually begin with, "But you just don't understand," and then continue on with, "We need to have the slides to send ahead!" Or "We need to provide leave-behinds!" Or "My audience wants the details!" Or "The slides help me remember what to say!" Or "My board member made me do it!" Or, the most common, "The slides have to stand alone!"

As a coach, I have tried to counter these objections by driving a wedge between the display functions (during the presentation) of slides and their document functions (before and after the presentation), to no avail. Despite all my efforts—and those of countless other fellow coaches, critics, and authors—presenters continue to treat presentations as documents.

So let's try one other argument. To set the stage, let me begin by describing the span of my services. For more than two decades, I have coached:

- Individual presenters to deliver one presentation to one audience, as in a keynote speech
- Individual presenters to deliver one presentation to multiple audiences, as in an IPO road show
- Individual presenters to deliver multiple presentations to multiple audiences, as in a tour to launch of a family of products
- Multiple presenters to deliver one presentation to one audience, as in presenting different levels of expertise at a conference or convention

- Multiple presenters to deliver multiple presentations to different audiences, as in offering a variety of schedule options at a conference or convention

But I had never coached multiple presenters to deliver one presentation to multiple audiences, until I worked with a unique group of executives at Cisco Systems.

Cisco has been my client ever since I coached its IPO road show in 1990, and they have given me the opportunity to work with many different business units. One of them, the Eastern Europe region, is run by Kaan Terzioglu, a Cisco vice president. Kaan invited me to Cisco's London facility to coach a dozen of his managers from several countries, including Croatia, Hungary, Poland, Bulgaria, Romania, and Turkey. Because of time, travel, and cost considerations, we reconfigured our core program, which usually involves only 4 participants, to accommodate all 12 people within one week. This format provided the identical set of techniques we give to the smaller groups, but it condensed some of the individual coaching. As a result, instead of developing 12 different presentations, we worked with only one. That one came from a set of slides that had originated in Cisco's corporate headquarters in San Jose.

In the culminating session of the program, I asked each participant to deliver a version of the slide show to an intended audience in his or her country. Their selections were quite diverse. One manager targeted a corporate enterprise, another a government agency, another a telecommunications carrier, and another a university. Yet as each participant stood to present, each was able to customize his or her story to a specific audience while using the same set of slides.

Multiple presenters were able to deliver one presentation to the multiple audiences because each presenter's narrative added value beyond the content of the slides.

The presenter is the focus of the presentation.

# 77

## The Art and Science of Oprah Winfrey
### *The Secrets of Oprah Winfrey's Appeal*

Oprah Winfrey shares a unique distinction with Lucille Ball, the 1950s television comedienne, and Mary Pickford, the 1920s movie actress: All three attained extraordinary popularity in front of the camera, and all three became powerhouses behind the camera as heads of their own production studios. But Ms. Winfrey differs from the other two women, in that her stardom is based on her own personality rather than on the assumed role of a character in a comedy or drama.

Ms. Ball played Lucy, a scatterbrained housewife, in her television series, and Ms. Pickford, known as "America's Sweetheart," portrayed ingenue leads in her films. Ms. Winfrey, on the other hand, from the very start of her career in television news and through every progressive step along her way to her own enormously successful *The Oprah Winfrey Show*, has always been herself—just Oprah. Her uncanny ability to be natural in all settings has enabled her to create, as her own web site states, "an unparalleled connection with people around the world."

Very few people can attain Oprah's level of success, but you can learn to make unparalleled connections with your own audiences by analyzing and adopting the elements of her style. Her talent to connect is both an art and a science; the science is the foundation, and the art is the expression.

The science is *empathy*, the universal human dynamic that has recently been gaining attention in the media and scientific communities. Empathy, evolved from the Greek word for emotion or affection, refers to shared or vicarious feelings, as distinct from *sympathy*, which is more about pity and implies separate feelings. In presentations, empathy is a sharing of feelings between presenters and audiences.

On Oprah's show, the empathy that resonates between her and her guests—whether they are celebrities or men and women from ordinary walks of life—fairly jumps off the video screen. Oprah clearly understands the pain and pleasure—the entire gamut of her guests' emotions—that she shares with her audiences. That sharing produces a cycle of emotions that generates further empathy in her audiences.

To see how she communicates the empathy, the art that leverages the science, let's compare her style with that of other prominent talk show hosts. We'll focus on seven key presentation factors. Although each of these hosts is quite successful in his own right, none has nearly the emotional impact on his audiences that Oprah does on hers.

1. **Roles.** Because of her grounding in news, Oprah conducts her interviews by immersing herself in her subjects' stories. David Letterman, Jay Leno, and their contemporaries Jon Stewart, Stephen Colbert, Jimmy Kimmel, Jimmy Fallon, Craig Ferguson, and Bill Maher, as well as the illustrious predecessor of them all, Johnny Carson, began their careers as comedians and so, during their interviews, function as performers. Oprah assumes the role of a congenial conversationalist with her guests. Other talk show hosts strive to match or outdo their guests' stories. Only Larry King, with his origins as a newscaster, gave his guests their full due during his 25 years on the air.

2. **Interaction.** Oprah listens carefully to her guests and responds warmly to their stories. The TV listings called Larry King "avuncular," which described his affability but set him apart from his guests. The comedians who go for the laughs widen that gap.

3. **Eye contact.** Oprah spends most of her air time engaged directly with her guests, making eye contact. Her counterparts, because of their performance orientation, play to their studio audience or to the camera and, therefore, to the vast unseen universe of viewers, appearing glib but impersonal. Larry King was the one exception among the others; he spent most of his air time in eye contact with his guests. Eye contact creates sincerity; sincerity generates empathy.

4. **Setting.** Oprah sits on a comfortable upholstered chair facing her guests, with nothing but air between them. Most other talk show hosts, including Larry King, sit behind a desk, the perennial standard of talk show decor. A desk on a talk show is the equivalent of a lectern in a speech: a barrier that diminishes empathy.

5. **Posture.** Oprah sits relaxed and open in her chair. The desks force the other talk show hosts to either sit up ramrod straight or slouch on the desktop.

6. **Gestures.** Oprah rarely uses props, leaving her hands free to gesture expressively and expansively. Other talk show hosts handle coffee cups, pencils, pens, index cards, and photographs, which often lead to distracting mannerisms.

7. **Smiles.** Many of Oprah's guests are the recipients of her generosity or the generosity of her sponsors. These "makeover" episodes produce smiles from the guests and Oprah smiles along empathically, radiating warmth both ways. Most of the other talk show hosts, observing the venerable show business rule of never laughing at one's own jokes, play deadpan (except for Jon Stewart, who, as an actor and a comedian, is a man of many funny faces).

To paraphrase Stephen Covey, Oprah demonstrates seven habits of a highly effective person—and a television superstar. To apply Oprah's seven habits to your presentations:

1. **Be conversational.** Follow the advice you've read several times throughout this book: Treat every presentation as a series of person-to-person conversations.

2. **Interact.** Read your audience as your presentation progresses, and be prepared to pause and adjust your content to keep them engaged. If you see disengagement or doubt, explain what you are saying or ask whether there is a question.

3. **Make eye contact.** As you proceed with your person-to-person conversations, look at each person until you see him or her look back at you.

4. **Present seated.** As you read in Chapter 52, "Presentation Advice from Titian," present at eye level whenever you can, depending on the size of the audience and the sight lines. Being at eye level creates empathy and re-creates the conversational setting. A general rule of thumb for presenting seated or standing is ten people: You can usually see every person in a seated group of nine; more than that requires you to stand to be able to make eye contact with everyone.

5. **Posture.** Whether you present seated or standing, be sure that your posture is straight. One way to check is to try to make your shoulder blades touch several times during your presentation. This simple technique will elevate your head and chin and make you appear poised. Try it and feel it.

6. ***Gesture.*** Use gestures to illustrate your words, but don't choreo-graph them. Do what comes naturally.

7. ***Smile.*** "When You're Smiling, the Whole World Smiles with You" is an old song (F77.1) recorded first by Louis Armstrong in 1929 and since then by countless other singers. The lyrics are just as applicable today because they identify empathy, the science behind Ms. Winfrey's art. Empathy is the sharing of feelings between presenters and audiences. If audiences see tentative or nervous behavior, they become dubious of the presenter; if they see confident or assertive behavior, they become trusting. Smiling produces a positive perception.

Practice these seven habits to become a highly effective presenter.

# 78

## Right or Left
### *The Deep Roots of Human Preferences*

Olivier Fontana of Microsoft Corporation, whose email about fast talking you read in Chapter 51, "Fast Talking," also wrote about how right and left preferences affect presentations. His email was prompted by a *Newsweek* article that reported on a scientific study of personal preferences driven by handedness:

> *[P]eople with different physical characteristics, such as being right- or left-handed, form different abstract concepts, corresponding to those physical traits. For southpaws, the left side of any space has positive moral, intellectual, and emotional connotations; for righties, the right side does. (F78.1)*

We live in a right-dominant world. Estimates of the right-handed majority range from five to one all the way up to nine to one. This dominance is also reflected in our language; think about the many common phrases that attribute positive values to the right:

- "It's all right with me"
- "All's right with the world"
- "My right hand"
- "Right-of-way"

Conversely, think about the many common phrases that attribute negativity to the left:

- "Left out"
- "Out in left field"
- "Two left feet."
- "Left-handed compliment"

The Latin origins of the words—*dexter* means "right" and *sinister* means "left"—carry the same values forward. The French counterpart extends them still further: *Gauche* means not only "left," but also "wrong." Coming full circle, *gauche* is now part of our English vocabulary meaning "lacking social polish" and "tactless."

According to some theories, the roots of this division of values go back to our distant ancestors. One comes from Rudolf Arnheim, the author of *Art and Visual Perception*, a 1954 book that is a bible for cinema students because of its theories governing camera movement. With regard to the left-to-right preference, Mr. Arnheim hypothesized that early humans were influenced by the sun's movement across the sky from left to right. (F78.2)

Another theory comes from the web site www.bigsiteofamazingfacts. com:

> *A person who was born right-handed would fight with a weapon in his right hand and use his left hand to shield himself; a left-handed person would fight with his left hand and shield himself with his right hand. A person who uses his left hand to shield himself protects his heart, which is on the left side of the body. So, many right-handed persons who were wounded would survive, while left-handed persons would suffer wounds around their heart and die.*
>
> *Over the course of evolution, this higher survival rate among right-handed persons could have led to more persons being born right-handed. (F78.3)*

Whether or not these theories are valid, we have evolved into a right-dominant world—Mother Nature at work.

But nurture is involved, too. In Western culture, we have all learned to read text from left to right. As a result, all movement to the right is more natural and, therefore, more appealing than to the left. This basic imprinting is so potent that it influences all people, both righties and lefties.

The roots of text direction also trace back to our early ancestors, when ancient writing was done on stone with a hammer and chisel. A right-handed person held the hammer in the right hand, the chisel in the left, and wrote right to left, to be able to see the letters forming. Therefore, ancient Hebrew and Arabic text, coming from the Stone Age, reads right to left. When paper and ink came into use, a right-handed person trying to write right to left, would smudge the wet ink; so in newer languages, the direction of text switched. (F78.4)

In Western cultures today, reading from left to right is so deeply embedded in childhood that it becomes second nature in adulthood. The innate predisposition of the eyes to move toward the right is irresistible. You can feel it as you scan this very page or the hard or soft copy of pages of any book, magazine, newspaper, or web site. Try moving your eyes the opposite way from right to left, and you'll feel a resistance.

Video and cinema directors incorporate this dynamic in how they direct their subjects and cameras. Watch a well-directed television drama or film and notice how the characters move across the screen. Most often, the sympathetic characters, the heroes and heroines, move from the left side of the screen toward the right, flowing with the natural movement of the eyes. By contrast, the unsympathetic characters, the villains, move from right to left, fighting the eyes' natural flow.

All these dynamics add up to a significant factor in presentations, with particular regard to the position, movement, and direction of all matters visual. This includes the design and animation of your graphics, and even the positioning of the physical elements of your presentation—as well as you. Whenever you present, put the projection screen to your left as you face the audience:

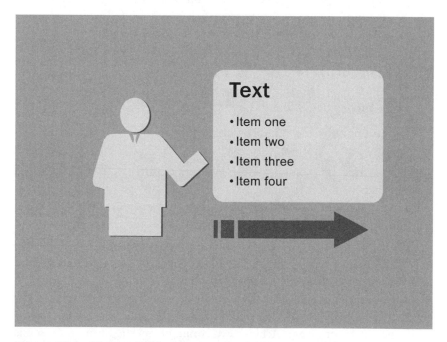

**Figure 78.1   The projection screen should be to your left.**

In this arrangement, every time you click to a new slide, the eyes of your audience will travel from you to the screen and across the image easily and naturally. This is especially important with text slides, so that the audience takes in the words on the screen just as they do in a book. If the screen were on the opposite side, the eyes of your audience would have to go backward, against the grain, before returning to take in the words on a second pass.

This same dynamic is applicable to how you animate your slides. Make the default entry movement of *all* your graphics from left to right, unless you want to send a negative or different message.

You, your slides, and the screen are all subject to these deeply embedded instinctive and learned forces acting upon your audience. When you step up to the front of the room, be sure you make the *right* choices.

# 79

## Graphics Synchronization
### *The Missing Link*

Mark Twain's nineteenth-century adage, "Everybody talks about the weather, but nobody does anything about it," is applicable to twenty-first-century presentations. In business today, everybody talks about Microsoft PowerPoint, the medium of choice for presentations. Most of the talk is about design; how to avoid making a visual hindrance of what is supposed to be a visual aid; and how to avoid the all-too-common "Death by PowerPoint." Multiple Amazon listings, abundant bookstore shelves, countless web sites, and numerous state-of-the-art graphics studios are all bursting at the seams with advice on how to design slides for presentations.

Yet nobody is doing anything about the other vital element meant to complement the graphics: the presenter. Oh, yes, advice about body language abounds, but there's nothing about how to *integrate* body language with the slides and the narrative.

This missing link creates a distraction during presentations that is as disconcerting as watching a film with an out-of-sync soundtrack. The movie audience, irritated by even the slightest mismatch of picture and sound, is likely to call out to the projectionist or even to ask for a refund. The business audience, struggling to relate what they are seeing with what the presenter is saying, is likely to interrupt or simply tune out, rejecting both the presenter and the message.

Such negative reactions occur because asynchronous sights and sounds challenge the sensitive neurology of the human perception system. Audiences find it difficult to process multiple sensory inputs, a difficulty compounded when the images are in motion—thus the irritation caused by the slipped soundtrack.

The equivalent of motion in presentations is the animation feature in PowerPoint. We've all been victimized by the flying bullets and spinning pie charts that tumble helter-skelter onto the projection screen like circus acrobats. In Chapter 38, "Computer Animation," you read how to exercise restraint when you use animation in your presentation. For now, let us accept that well-designed animation can help tell and propel a story, and turn our attention to how the presenter can incorporate animation into a presentation effectively.

At the instant the animation begins, the audience shifts attention to the screen and *away* from the presenter *involuntarily*—that sensitive neurology at work. The audience is so focused is on the animation, they do not hear the presenter's words, nor do they see what the presenter is doing. Moreover, *anything* that the presenter does or says creates additional sensory data that conflicts with the projected activity on the screen.

There is a simple solution to all of this: Pause. Whenever you introduce animation, stop talking, stop moving, turn to the screen, and let the animation complete its full course of action. In fact, whenever you introduce *any* new graphical element, even a static image, pause and look at it. Look at the image as if you've never seen it, and give your audience time to see it. At that moment, you and your audience fall into step.

The pause is the centerpiece of Graphics Synchronization, a unique skill set that integrates the presenter's delivery and narrative with both the design and the animation of the graphics. You can read more about this skill in *The Power Presenter;* for now, let's focus on the value of the pause.

- You get to look at your slide to make sure that it's correct.
- You get a prompt about what to say.
- You get to take a breath and keep living.

One more benefit is more important than all the others:

- Your audience gets time to absorb your slide and get a visual reinforcement of your message.

You get all these benefits for the price of doing absolutely nothing. Think about that: The key to integrating all the critical elements of a presentation is not what to do; it is what *not* to do.

Pause.

# 80

## The House That Jack Built
### *Make All the Parts Fit*

This is the house that Jack built.

This is the rat

That ate the malt

That lay in the house that Jack built.

This is the cat,

That killed the rat,

That ate the malt

That lay in the house that Jack built.

This is the dog,

That worried the cat,

That killed the rat,

That ate the malt

That lay in the house that Jack built.

The repetitive progression structure of this classic Mother Goose nursery rhyme has been applied in many variations with many stories in many cultures. In all its expressions, the progression shows the continuous interrelationship among disparate components. The theme is also applicable to presentations in the interrelationships among the key components of every presentation:

- Content
- Graphics
- Delivery skills
- Q&A

Suppose that a presenter developed a clear, concise, and compelling story. But then suppose that the presenter accompanied that story with graphics designed in the "Death by PowerPoint" manner. The story would fail.

Suppose instead that a presenter developed a clear, concise, and compelling story, and accompanied that story with graphics designed in the *Less Is More* manner. But then suppose that the presenter stood up in front of the audience and suddenly froze like a deer in the headlights. The story would fail.

Suppose now that a presenter developed a clear, concise, and compelling story; accompanied that story with graphics designed in the *Less Is More* manner; and then stood up in front of the audience and delivered the presentation with the outstanding oratorical skills of Ronald Reagan or Barack Obama. The combination was so impressive that the audience sat in awed silence for the entire length of the presentation, not uttering a peep. But then suppose that, when the presenter concluded the presentation and opened the floor to questions, the first question was hostile and the presenter reacted defensively. Despite everything that preceded, the entire presentation would fail.

The point here is that, for any presentation to succeed, every presenter must give full attention to every component. More to the point, the presenter must be certain that each component integrates with every other component.

Build your house better than Jack did.

# Footnotes

F1.1 *The Wizard of Oz* (film directed by Victor Fleming 1939)

F1.2 http://www.newseum.org/todaysfrontpages/flash/

F2.1 http://www.time.com/time/covers/0,16641,20061225,00.html

F2.2 http://www.google.com/search?sourceid=navclient&aq=t&ie=
UTF-8&rls=GGLG,GGLG:2005-34,GGLG:en&q=yale+
university+persuasive+words

F2.3 http://www.newyorker.com/reporting/2007/11/26/
071126fa_fact_lizza

F2.4 http://www.nytimes.com/2009/01/20/us/politics/20text-obama.
html?_r=1

F2.5 http://www.nytimes.com/2009/01/21/us/politics/21kakutani.html?
scp=8&sq=&st=nyt

F5.1 http://www.smithsonianmag.com/history-archaeology/man-of-
his-words.html

F7.1 http://www.twainquotes.com/

F8.1 http://www.nytimes.com/2009/04/12/movies/12mcgr.html?_r=1&
scp=2&sq=mike%20nichols&st=cse

F9.1 http://www.sfgate.com/cgi-bin/article.cgi?f=/c/a/2008/08/15/
DD62124RD6.DTL&hw=vicky+cristina+barcelona+lasalle&sn=
001&sc=1000

F11.1 R.W. Apple Jr., "The Democrats in Atlanta; Dukakis' Speech
Offers His 'Vision of America'," *New York Times*, July 22, 1988

F11.2 Bill Clinton, *My Life* (New York: Random House, Inc., 2004)

F12.1 http://online.wsj.com/article/SB124596573543456401.html

F12.2 http://www.thedailyshow.com/full-episodes/232256/thu-july-2-
2009-robert-kenner

F12.3 http://adage.com/century/people005.html

F13.1 Robert Green, *The Laws of Power* (Viking Adult, 1998)

F14.1 http://online.wsj.com/article/SB124535297048828601.html

F14.2 http://www.nytimes.com/2009/07/05/magazine/05FOB-OnLanguage-t.html

F17.1 http://mashable.com/2010/02/01/condensed-ipad-keynote/

F18.1 http://online.wsj.com/article/SB10001424052748703740004574513463106012106.html#mod=todays_us_weekend_journal

F18.2 http://john-irving.com

F19.1 http://online.wsj.com/article/SB100014240527487037400045745133463106012106.html#mod=todays_us_weekend_journal

F20.1 http://www.nytimes.com/2002/11/24/magazine/microsofter.html?scp=1&sq=&st=nyt

F21.1 http://www.nejm.org/doi/full/10.1056/NEJM199303183281123

F22.1 http://www.nytimes.com/2010/04/25/books/review/Halperin-t.html?ref=books

F22.2 http://thecaucus.blogs.nytimes.com/2010/04/25/after-15-months-in-office-policy-vs-politics-for-obama/?ref=todayspaper

F26.1 Edward Tufte, *The Cognizant Style of PowerPoint* (Graphics Pr, 2003)

F29.1 Dr. Stephen Kosslyn, *Clear and to the Point: 8 Psychological Principles for Compelling PowerPoint Presentations* (Oxford University Press, 2007)

F29.2 http://www.wjh.harvard.edu/~kwn/Kosslyn_pdfs/2005Chabris_chap_in_KnowledgeInfoVisualization_RepresentationalCorrespondence.pdf

F30.1 http://office.microsoft.com/en-us/templates/TC103382711033.aspx?CategoryID=CT103366151033

F32.1 http://www.youtube.com/watch?v=GB7S-KOJIfE

F33.1 http://graphics8.nytimes.com/packages/pdf/weekinreview/20090531_corporate_logos.pdf

F35.1 http://www.navy.mil/navydata/ships/carriers/rainbow.asp

F39.1 http://www.nytimes.com/2010/04/27/world/27powerpoint.html? hp

F39.2 http://worldblog.msnbc.msn.com/_news/2009/12/02/4376696-so-what-is-the-actual-surge-strategy

F39.3 http://www.afji.com/2009/07/4061641

F40.1 Catherine Blyth, *The Art of Conversation: A Guided Tour of a Neglected Pleasure* (Gotham, 2008)

F40.2 Marshall H. Klaus, John H. Kennell, and Phyllis H. Klaus, *Bonding: Building the Foundations of Secure Attachment and Independence* (Da Capo Press, 1996)

F42.1 http://online.wsj.com/article/ SB10001424052970203863204574348493483201758.html

F43.1 http://online.wsj.com/article/SB123802014251941619.html

F44.1 http://www.kipling.org.uk/poems_if.htm

F44.2 http://online.wsj.com/article/SB124424873407590721.html

F45.1 http://www.nytimes.com/2009/07/26/sports/26swim.html?scp=4 &sq=karen%20crouse&st=cse

F49.1 http://www.washingtonpost.com/wp-dyn/content/article/2007/02/ 22/AR2007022202189.html

F51.1 http://www.youtube.com/watch?v=NeK5ZjtpO-M

F53.1 http://online.wsj.com/article/ SB10001424052748704576204574529832100929484.html

F54.1 http://online.wsj.com/article/ SB10001424052748703298004574454973666650080.html

F55.1 *Gypsy* (film directed by Mervyn LeRoy, 1962)

F55.2 http://online.wsj.com/article/SB1000142405274870380790457500 97390662423192.html?KEYWORDS=teachout

F57.1 http://www.youtube.com/watch?v=M4qGKdl37iY&feature=Play List&p=2AA13CABCBB3F96C&index=4&playnext=2&playnext _from=PL

F57.2 http://www.telegraph.co.uk/news/uknews/1564018/Er-a-little- hesitation-can-um-help-your-brain.html

F58.1 http://www.rollingstone.com/politics/news/17390/119236

F58.2 http://content.usatoday.com/communities/greenhouse/post/2010/ 06/bp-tony-hayward-apology/1

F58.3 http://www.washingtonpost.com/wp-dyn/content/article/ 2006/08/14/AR2006081401114.html

F58.4 http://www.washingtonpost.com/ac2/wp-dyn?pagename= article&contentId=A20730-2002Dec6&notFound=true

F58.5 http://online.wsj.com/article/SB1000142405274870417800457530 50940170440292.html?mod=ITP_personaljournal_0

F61.1 Jean Anthelme Brillat-Savarin, *The Physiology of Taste: or Meditations on Transcendental Gastronomy* (Everyman's Library, 2009)

F61.2 Michael Pollan, *Food Rules: An Eater's Manual* (Penguin, 2009)

F63.1 http://www.nytimes.com/2009/04/29/us/politics/29text-obama.html?ref=politics

F65.1 http://online.wsj.com/article/SB123361001775640785.html

F67.1 Courtesy CNN

F67.2 http://www.nytimes.com/2009/02/09/us/politics/09text-obama.html

F68.1 http://www.thedailyshow.com/full-episodes/index.jhtml?episodeId=218354

F68.2 http://www.usatoday.com/money/media/2009-03-11-cnbc-cramer-stewart_N.htm

F68.3 http://www.businessweek.com/careers/managementiq/archives/2009/03/jon_stewart_vs.html

F69.1 David F. Bellet, Retired Chairman Crown Advisors

F70.1 http://www.nfl.com/players/jerryrice/careerstats?id=RIC128880

F71.1 http://abcnews.go.com/ThisWeek/week-transcript-kaine-steele/story?id=10721384&page=2

F71.2 http://www.nytimes.com/2010/05/24/opinion/24douthat.html?hp

F72.1 http://www.realclearpolitics.com/news/ap/politics/2010/Jun/23/kagan_practices_answers__poise_in_mock_hearings.html

F73.1 http://www.thedailyshow.com/watch/thu-july-1-2010/release-the-kagan

F73.2 http://opinionator.blogs.nytimes.com/2010/07/01/past-present-and-future-justice/?emc=eta1

F74.1 http://rack1.ul.cs.cmu.edu/is/saxe/doc.scn?fr=0&rp=http%3A%2F%2Frack1.ul.cs.cmu.edu%2Fis%2Fsaxe%2F&pg=4

F78.1 Reprinted by permission of Olivier Fontana

F78.2 Rudolf Arnheim, *Art and Visual Perception: A Psychology of the Creative Eye* (University of California Press, 2004)

F78.3 http://www.bigsiteofamazingfacts.com/why-are-more-people-right-handed-than-left-handed

F78.4 Courtesy of Eyal Waldman, Chairman of the Board and Chief Executive Officer of Mellanox Technologies

# Acknowledgments

My sincere gratitude for their contributions to Power Presentations, Ltd., and the programs from which this book is derived goes out to: Nichole Nears, Pearl Cheung, Jim Welch, Jon Bromberg, Rich Hall, Brad Seals, and David Weissman (no relative) and Bob Johns of Video Arts.

For their contributions to the publication of this book: my attorney, Bill Immerman; my agent, Jim Levine and Kerry Sparks of the Levine Greenberg Literary Agency; my Executive Editor at Pearson Jeanne Glasser, and her colleagues, Lori Lyons, project editor, Krista Hansing, copy editor, Gloria Schurick, compositor; and, again, Nichole Nears, for managing the manuscript and graphics.

For their contributions to the stories in this book: Olivier Fontana, John Boladian, and Matt Vasey, Microsoft Corporation; Kaan Terzioglu, Cisco Systems; Jason Trujillo, Intel Corporation; Mike Wallace, CBS News; Bill Portelli, CollabNet; Bill Peacock, Commonwealth Club; James E. Muller, M.D., InfraReDx; Stephen Kosslyn, Ph.D., Harvard University; Marshall H. Klaus, M.D.; Dave Vieau and Mike Rubino, A123 Systems; David Bellet, Crown Advisors International; Gino Giglio, Association of Theatrical Press Agents and Managers; and Eyal Waldman, Mellanox Technologies

For their support and love: my kids, Bixby and the AR Emeritus, and Natalie.

# About the Author

**Jerry Weissman** is the world's number one corporate presentations coach. His private client list reads like a who's who of the world's best companies, including the top brass at Cisco Systems, Microsoft, Yahoo!, Intel, Intuit, Netflix, Dolby Labs, and many others.

Mr. Weissman founded Power Presentations, Ltd. in 1988. One of his earliest efforts was the Cisco Systems IPO road show. Following its successful launch, Don Valentine, of Sequoia Capital, and then chairman of Cisco's Board of Directors, attributed "at least two to three dollars" of the offering price to Mr. Weissman's coaching. That endorsement led to more than 500 other IPO road show presentations that have raised hundreds of billions of dollars in the stock market.

Mr. Weissman's focus widened from coaching IPOs to include public and privately held companies. His techniques have helped another 500 plus firms develop and deliver their mission-critical business presentations.

Mr. Weissman is also the author of three books, the bestselling *Presenting to Win: The Art of Telling Your Story*, named by *Fortune* magazine as one of eight must-reads; *The Power Presenter: Technique, Style, and Strategy from America's Top Speaking Coach*; and *In the Line of Fire: How to Handle Tough Questions...When it Counts*.

# INDEX

## A

A123 Systems, 94
accountability when answering questions, 136-137
Active Listening
   to multiple questions, 133-134
   silent reactions and, 135
   speed of Q&A responses, 131-132
   understanding questions before answering, 146
"Aha!" Moments, 32-33
airline example (flow of presentations), 34-35
Alito, Samuel, 149
all-caps, initial caps versus, 71-72
Allen, George, 122
Allen, Woody, 22
anecdotes. *See* human interest stories
animation
   pausing, 168-169
   of text, 80-81
   tips for, 82-83
answers, keeping short, 144-145. *See also* Q&A sessions
Archimedes, 32
Aristotle, 17, 41
Armstrong, Louis, 163

Army (U.S.), 24-25
Arnheim, Rudolf, 165
*Art and Visual Perception* (Arnheim), 165
*The Art of Conversation* (Blyth), 89
asking questions, customizing presentations, 6
Astaire, Fred, 101, 113
audience
   assuming intelligence of, 36-37
   avoiding disconnect with, 10-11
   benefits for, 44-45
   concentration on, 95-98
   connecting with, 8-9, 24-25, 160-163
Audience Advocacy, 46-49

## B

Ball, Lucille, 160
Ballmer, Steve, 45
Barber, Red, 115
basketball example, 124-125
"Be All That You Can Be" slogan, 24-25
Bell, Alexander Graham, 32
Bellet, David, 144
benefit statements. *See* WIIFY (What's In It For You?)

benefits for audience, 44-45
Blanchett, Cate, 22
Blumenthal, Richard, 147
Blyth, Catherine, 89
Bodow, Steve, 45
body language, 118-119
"Body Wrap" position, 118-119
Bohan, Caren, 141
*Bonding: Building the Foundations of Secure Attachment and Independence* (Klaus, Kennell, Klaus), 89
Brand, Rachel, 149
breath, pausing for, 103-104
breathing, 121
brevity in answering questions, 144-145. *See also* length of presentations; slogans
Brillat-Savarin, Jean Anthelme, 127
bullets, clarity of, 67-68
Bumiller, Elisabeth, 84
Bush, George H.W., 9, 50
Bush, George W., 48, 50, 149

## C

cadence of voice, 101-102
call to action, 44-45
capitalization, all-caps versus initial caps, 71-72
Carson, Johnny, 12, 135, 161
Cervantes, Miguel, 38
Chaucer, Geoffrey, 32
Chronological Flow Structure, 17
Cisco Systems, 158-159
clarity, 15, 67-68
*Clear and to the Point: 8 Psychological Principles for Compelling PowerPoint Presentations* (Kosslyn), 62

Clinton, Bill, 26, 50, 106-107
Clooney, George, 91
CNN, 80, 126
*The Cognitive Style of PowerPoint* (Tufte), 57
Colbert, Stephen, 161
color in graphics, 75, 77
combining pictures and text, 64-66
communication in sales, 60-61
complex slides, usefulness of, 84-85
concentration on audience reaction, 95-98
confidence when presenting, 113-114
connection with audience, 8-9, 24-25, 160-163
contemporization, customizing presentations, 6
contextual perception, 155
contrast colors in graphics, 75, 77
conversational style, 89-90, 93-94, 161
    Bill Clinton, 106-107
    Edward R. Murrow, 91-92
    Elizabeth Dole, 108
    Vin Scully, 115
Cooper, Anderson, 58
Corona, 78-79
*Courage and Consequence: My Life as a Conservative in the Fight* (Rove), 48
Couric, Katie, 58
Covey, Stephen, 17, 41, 162
Cramer, Jim, 142-143
*The Curious Case of Benjamin Button* (film), 22
Curtis, Neil, 38
customizing presentations, 5-7

# D

*The Daily Show* (television program), 135, 142
Danticat, Edwidge, 42-43
Davis, Julie Hirschfeld, 149
delivery style
  body language, 118-119
  concentration on audience reaction, 95-98
  confidence in, 113-114
  eye contact, 89-90, 105
  first impressions, 124-125
  food and drink when speaking, 127
  group dynamics, 106-107
  nervousness, overcoming, 99-100, 116-117
  pausing for breath, 103-104
  person-to-person conversation, 91-94, 108, 115
  position (sitting versus standing), 111-112
  speaking cadence, 101-102
  speed of speaking, 109
  tips for, 126
  unwords, 120-121
  verbal gaffes, 122-123
DeNiro, Robert, 138
Dershowitz, Alan, 142
DiMaggio, Joe, 113
direct references, customizing presentations, 6
disconnect with audience, avoiding, 10-11
documents, presentations versus, 55-58, 158-159
Dodd, Christopher, 147
Dole, Elizabeth, 108
Douthat, Ross, 147
drafts of presentations, 40-41

drinks when speaking, 127
Duarte, Nancy, 85
Dukakis, Michael, 26

# E

elephant poem example, 155
elevator pitches, 28-29
empathy, 160-163
Engel, Richard, 84
Epictetus, 132
Erasmus, 105
exaggeration, avoiding, 38-39
eye contact, 89-90, 105-107, 161

# F

Fallon, Jimmy, 161
false assumption questions, answering, 140-141
fast talking, 109
fear of public speaking, overcoming, 95-100, 116-117
Feldshuh, Tovah, 97
Ferguson, Craig, 161
Fields, Dorothy, 101
"Fig Leaf" position, 118-119
filler words, 120-121
Fincher, David, 22
first impressions, 124-125
first slide, customizing presentations, 7
flow of presentations, 34-35
Flow Structures of presentations, 16-17
Fontana, Olivier, 109, 164
fonts, serif versus sans serif, 73-74
food and drink when speaking, 127
*The 48 Laws of Power* (Greene), 31

free throw line example, 124-125
frequently asked questions,
    answering, 136-137

## G

Gallo, Carmine, 38
Geithner, Timothy, 118
gestures, 119, 163
*The Godfather, Part II* (film), 49
*Good Night, and Good Luck*
    (film), 91
Goodman, Benny, 116
Gould, Glenn, 116
graphics
    animation tips, 82-83
    clarity of text, 67-68
    color in, 75, 77
    headlines versus sentences, 59
    initial caps versus all-caps, 71-72
    picture/text combinations, 64-66
    serif versus sans serif fonts,
        73-74
    storytelling versus, 55-58,
    158-159
    synchronizing with linguistics,
        156-157, 168-169
    text animation, 80-81
Graphics Synchronization
    skills, 169
Greene, Robert, 31
Greenhouse, Linda, 151
group dynamics, 106-107
*Gypsy* (musical), 116

## H

Halperin, Mark, 48
*Hamlet* (Shakespeare), 144
Hammes, Thomas X., 85

handedness, 164-167
Hardy, Oliver, 135
Hayward, Tony, 122
headline style, sentence style
    versus, 59, 71-72
Heifetz, Jascha, 113
Heilemann, John, 48
Hoffman, Dustin, 138
Horowitz, Vladimir, 95
"How NOT to Use PowerPoint"
    (McMillan), 69
human interest stories, 50-52
humor in presentations, 12-13
hyperbole, avoiding, 38-39

## I

"If" (Kipling), 97
inferior angle, 111
initial caps, all-caps versus, 71-72
integration
    contextual perception and, 155
    graphics with linguistics,
        156-157
    of presentation components,
        170-171
intelligence of audience,
    assuming, 36-37
interpersonal communication. *See*
    person-to-person conversation
interviews, tips for, 126
intuitiveness of presentations,
    36-37
iPhone, 36-37
IPO road shows, 94
    length of, 27
*Irena's Vow* (Broadway play), 97
Irving, John, 40-41

## J–K

job interviews, tips for, 126
Jobs, Steve, 38-39
jokes, avoiding, 12-13

Kagan, Elena, 149-152
Kaine, Tim, 147
"Keeping Your Foot Away from Your Mouth" (Zaslow), 123
Kennedy, John F., 14, 28, 105
Kennell, John H., 89
Kern, Jerome, 101
Kimmel, Jimmy, 161
King, Larry, 161
Kipling, Rudyard, 56, 97
Klaus, Marshall H., 89
Klaus, Phyllis H., 89
Kosslyn, Stephen, 62-63, 66
Kounios, John, 32

## L

Lane, Robert, 62
LaSalle, Mick, 22
Laurel, Stan, 135
left versus right, 164-167
length of presentations, 18-19, 26-27
Leno, Jay, 161
Letterman, David, 17, 120, 161
Lincoln, Abraham, 14
linguistics, synchronizing graphics with, 156-157, 168-169
listening
    to multiple questions, 133-134
    silent reactions and, 135
    speed of Q&A responses, 131-132
    understanding questions before answering, 146

localization, customizing presentations, 6
Lott, Trent, 122
Loven, Jennifer, 141
Lublin, Joann, 136
Luce, Clare Boothe, 28

## M

*Mad Money* (television program), 143
Madoff, Bernie, 142-143
Maher, Bill, 161
Mattis, James N., 84
McChrystal, Stanley A., 84, 122
McMaster, H. R., 84
McMillan, Don, 69
Mental Method of Presenting, 96
Microsoft, 44-45
Mies van der Rohe, Ludwig, 27
military, PowerPoint usage, 84-85
*Milk* (film), 22
"Miss Otis Regrets" (Porter), 151
mistakes
    admitting when questioned, 142-143
    during speaking, 122-123
Morgan, Frank, 5
Moschitta, John Jr., 109
Muller, James E., 46-47
multiple drafts of presentations, 40-41
multiple presenters to multiple audiences with single presentation, 158-159
multiple questions, listening to, 133-134
"Murder Boards" (preparation for Q&A sessions), 149-150
Murrow, Edward R., 91-92

Mutual of Omaha, 32
mutual references, customizing
   presentations, 6

# N–O

"naming things" process, 20-21
Navy (U.S.), 75
nervousness, overcoming, 99-100,
   116-117
*New York Times*, 80
Newton, Isaac, 32
Nichols, Mike, 20-21
Noonan, Peggy, 9, 28-29
Numerical Flow Structure, 17

Obama, Barack, 8-9, 26, 28-29,
   48, 50, 118, 120-122, 133-134,
   141, 149, 151
Olivier, Laurence, 116
*The Oprah Winfrey Show*
   (television program), 160
Orszag, Peter, 118
ownership, 15

# P

Paul, Rand, 147
pausing, 109
   to breathe, 103-104
   during presentations, 63, 78-79,
      168-169
Peabody, Francis W., 46
Peacock, Bill, 24, 108
Penn, Sean, 22
Perahia, Murray, 95-96
perception of presentations,
   62-63, 78-79, 164-167
person-to-person conversation,
   93-94, 161
   Bill Clinton, 106-107
   Edward R. Murrow, 91-92

Elizabeth Dole, 108
Vin Scully, 115
persuasion, power of "you," 8-9,
   44-45
Phelps, Michael, 99-100
phrases, positive phrases in
   presentations, 38-39
*The Physiology of Taste: Or,
   Meditations on Transcendental
   Gastronomy* (Brillat-
   Savarin, 127
Pickford, Mary, 160
pictures, combining with text,
   64-66
politicians, spinning the message,
   147-148
Pollan, Michael, 127
Polonius, 144
Pope, Alexander, 123
Portelli, Bill, 36
Porter, Cole, 151
position (sitting versus standing),
   111-112
positive phrases in presentations,
   38-39
posture, 162
Powell, Colin, 140
practicing presentations, 30-31,
   113-114, 123
preparation for presentations
   "naming things" process, 20-21
   time required, 18-19
preparation for Q&A sessions,
   149-150
*The Presentation Secrets of Steve
   Jobs: How to Be Insanely Great
   in Front of Any Audience*
   (Gallo), 38
*Presentation Zen: Simple Ideas on
   Presentation Design and
   Delivery* (Reynolds), 85

Presentation-as-Document
  Syndrome, 55-58
presentations
  contextual perception, 155
  customizing, 5-7
  delivery style. *See* delivery style
  documents versus, 55-58,
    158-159
  flow of, 34-35
  human perception and, 62-63,
    78-79, 164-167
  humor in, 12-13
  integration of components,
    170-171
  key factors in, 160-163
  length of, 18-19, 26-27
  multiple drafts of, 40-41
  pausing during, 63, 78-79,
    168-169
  practicing, 30-31, 113-114, 123
  preparation for
      *"naming things" process,*
      *20-21*
      *time required, 18-19*
  sending previews of, 60-61
  storytelling in, 22-23, 36-37
  structure of, 16-17
presidential speechwriting, 14
previews of presentation, sending,
  60-61
prompts, headline style, 59
public speaking, overcoming fear
  of, 95-100, 116-117
Publius Syrus, 31

# Q

Q&A sessions
  accountability in, 136-137
  admitting mistakes, 142-143
    false assumption questions,
      140-141
    multiple questions, 133-134
    preparation for, 149-150
    refusal to answer questions,
      151-152
    short answers, 144-145
    silent reactions, 135
    speed of responses during,
      131-132
    spinning the message, avoiding,
      147-148
    spinning versus Topspin,
      138-139
    understanding questions before
      answering, 146
questions, asking, customizing
  presentations, 6

# R

reading slides, 69-70
Reagan, Nancy, 50, 140
Reagan, Ronald, 50-51, 115
redirection, 99
Reeves, Rosser, 29
refusal to answer questions,
  151-152
rehearsing. *See* practicing
  presentations
Reid, Chip, 141
rewriting, 40-41
Reynolds, Garr, 85
Rice, Jerry, 146
Riddle, Nelson, 102
right versus left, 164-167
Robbins, Jerome, 116
Roberts, John G., 149
Romer, Christina, 118
Rove, Karl, 48-49
Rubino, Mike, 94

# S

Safire, William, 32
sales, as in-person
  communication, 60-61
sans serif fonts, serif fonts versus,
  73-74
Saxe, John Godfrey, 155
Scully, Vin, 115
sentence style, headline style
  versus, 59, 71-72
serif fonts, sans serif fonts versus,
  73-74
*The Seven Habits of Highly
  Effective People* (Covey), 41
Seward, William, 14
shading, slides, 67-68
Shakespeare, William, 144
short answers, 144-145
*Show Me! What Brain Research
  Says about Visuals in
  PowerPoint* (Kosslyn and
  Lane), 62
silent reactions when listening, 135
Simon, Carly, 116
simplicity of presentations, 36-37
Sinatra, Frank, 101-102
sitting versus standing,
  111-112, 162
Skelton, Red, 135
Skutnik, Lenny, 50
Slide Sorter view, 42-43
slides
  complex slides, usefulness of,
    84-85
  describing during presentation,
    156-157
  pausing while viewing, 63, 78-79,
    168-169

picture/text combinations, 64-66
  as presentation previews, 60-61
  reading, 69-70
*slide:ology* (Duarte), 85
slogans, 24-25, 28-29, 44-45
smiling, 163
"So What?" Syndrome, 10-11
Sondheim, Stephen, 116
Sorensen, Ted, 14
Sotomayor, Sonia, 149
spaced learning, 40-41
speaking cadence, 101-102
speaking speed, 109
speaking style. *See* delivery style
speed of Q&A responses, 131-132
speed of speech, 109
spinning the message
  avoiding, 147-148
  Topspin versus, 138-139
stage fright, 950
"Stand at Parade Rest"
  position, 119
standing versus sitting,
  111-112, 162
Stewart, Jon, 12, 28, 135,
  142-143, 151, 161-162
storyboarding, 42-43
storytelling
  graphics versus, 55-58, 158-159
  human interest stories, 50-52
  in presentations, 22-23, 36-37
structure of presentations, 16-17
structured relaxation, 100
Summers, Larry, 118
superior angle, 111
supporting data, customizing
  presentations, 6
sympathy, 160
synchronizing graphics and
  linguistics, 156-157, 168-169

# T

talking fast, 109
Tatum, Art, 113
Te Kanawa, Kiri, 103-104
Teachout, Terry, 113, 116
teleology, 41
television appearances, tips
  for, 126
terminology, positive phrases in
  presentations, 38-39
Terzioglu, Kaan, 159
text
  animating, 80-81
  clarity of, 67-68
  combining with pictures, 64-66
  inital caps versus all-caps, 71-72
  serif versus sans serif fonts,
    73-74
Thurmond, Strom, 122
*Time* (magazine), 8
time required for preparation,
  18-19
Titian, 111-112
Title *Plus* statements, 156-157
Toobin, Jeffrey, 16-17
Topspin, spinning the message
  versus, 138-139
Torres, Dara, 99-100
Trujillo, Jason, 31
Tufte, Edward, 57
Twain, Mark, 18-19, 168

# U–V

U.S. Army, 24-25
U.S. Navy, 75
understanding questions before
  answering, 146
Unique Selling Proposition
  (USP), 29
unwords, 120-121

value, adding, 15
Van Sant, Gus, 22
Vasey, Matt, 156-157
verbal gaffes, 122-123
Verbalization, 30-31, 42-43,
  113-114, 123, 149-150
*Vicky Cristina Barcelona*
  (film), 22
Vieau, Dave, 94
*The Visual Display of
  Quantitative Information*
  (Tufte), 57
visualization, 100

# W

*Wag the Dog* (film), 138
*Wall Street Journal*, 93
Wallace, Mike, 1
Warhol, Andy, 19
"The Way You Look Tonight," 101
"We Have Met the Enemy and
  He Is PowerPoint"
  (Bumiller), 84
Weil, Andrew, 16-17
Wells, Orson, 116
"What keeps you up at night?"
  question, answering, 136-137
"When You're Smiling, the Whole
  World Smiles with You"
  (Armstrong), 163
"Why Gen-Y Johnny Can't Read
  Nonverbal Cues" (*Wall Street
  Journal*), 93
*Why Men Don't Listen and
  Women Can't Read Maps: How
  We're Different and What to Do
  About It*, 79
WIIFY (What's In It For You?),
  10-11, 25

Williams, Brian, 58
Winfrey, Oprah, 32, 160-163
*The Wizard of Oz* (film), 5-7

## Y–Z

Yards After Catch (YAC), 146
"you," power of, 8-9, 44-45

Zaslow, Jeffrey, 123
Zeleny, Jeff, 133-134

FINANCIAL TIMES

In an increasingly competitive world, it is quality of thinking that gives an edge—an idea that opens new doors, a technique that solves a problem, or an insight that simply helps make sense of it all.

We work with leading authors in the various arenas of business and finance to bring cutting-edge thinking and best-learning practices to a global market.

It is our goal to create world-class print publications and electronic products that give readers knowledge and understanding that can then be applied, whether studying or at work.

To find out more about our business products, you can visit us at www.ftpress.com.